Forest Dynamics and Disturbance Regimes
Studies from Temperate Evergreen–Deciduous Forests

More than a quarter of the world's forests lie within the cool-to-cold temperate zones of the northern and southern hemisphere. Their distinctive mosaics of evergreen conifers and deciduous hardwood species have been shaped by fire, wind and herbivory over thousands of years. In the last few centuries human activities have increasingly changed the dynamics of these mosaics: fire suppression and exclusion have reduced fire frequency, harvesting and grazing have increased, and a changing climate may be leading to a change in the frequency of high winds. While the exact influence of these changes remains to be determined, this book provides a major contribution to the study of forest dynamics by considering three important themes:

- The combined influence of the three main disturbance types – wind, fire and herbivory – on the successional trajectories and structural characteristics of forests.
- The interaction of deciduous and evergreen tree species to form various mosaics which, in turn, differentially influence the environment and disturbance regime.
- The significance of temporal and spatial scale with regard to the overall impact of disturbances.

These themes are explored via case studies from the forests in the Lake States of the USA (Minnesota, Wisconsin and Michigan) where the presence of large primary forest remnants presents a unique opportunity to study and compare the long-term dynamics of near-boreal, pine and hemlock–hardwood forests. The comparability of these forests to forests in other cool-to-cold temperate zones allows generalizations to be made that may apply more widely.

LEE FRELICH is a Research Associate in the Department of Forest Resources at the University of Minnesota, St Paul and Founder and Director of the University of Minnesota Centre for Hardwood Ecology. His research experience spans studies of the impact of acid rain on forest growth, paleoecological studies of forest change, tree population dynamics, old-growth forest dynamics and forest disturbance ecology, including the effects of fire, windstorms and grazing.

T0275753

CAMBRIDGE STUDIES IN ECOLOGY

Editors
H. J. B. Birks   *Botanical Institute, University of Bergen, Norway, and Environmental Change Research Centre, University College London*
J. A. Wiens   *Colorado State University, USA*

Advisory Editorial Board
P. Adam   *University of New South Wales, Australia*
R. T. Paine   *University of Washington, Seattle, USA*
R. B. Root   *Cornell University, USA*
F. I. Woodward   *University of Sheffield, Sheffield, UK*

This series presents balanced, comprehensive, up-to-date, and critical reviews of selected topics within ecology, both botanical and zoological. The Series is aimed at advanced final-year undergraduates, graduate students, researchers, and university teachers, as well as ecologists in industry and government research.

It encompasses a wide range of approaches and spatial, temporal, and taxonomic scales in ecology, experimental, behavioural and evolutionary studies. The emphasis throughout is on ecology related to the real world of plants and animals in the field rather than on purely theoretical abstractions and mathematical models. Some books in the Series attempt to challenge existing ecological paradigms and present new concepts, empirical or theoretical models, and testable hypotheses. Others attempt to explore new approaches and present syntheses on topics of considerable importance ecologically which cut across the conventional but artificial boundaries within the science of ecology.

Forest Dynamics and Disturbance Regimes

Studies from Temperate Evergreen–Deciduous Forests

LEE E. FRELICH

University of Minnesota, Department of Forest Resources

CAMBRIDGE
UNIVERSITY PRESS

CAMBRIDGE UNIVERSITY PRESS
Cambridge, New York, Melbourne, Madrid, Cape Town, Singapore, São Paulo

Cambridge University Press
The Edinburgh Building, Cambridge CB2 8RU, UK

Published in the United States of America by Cambridge University Press, New York

www.cambridge.org
Information on this title: www.cambridge.org/9780521650823

First published 2002
Reprinted 2003
This digitally printed version 2008

A catalogue record for this publication is available from the British Library

Library of Congress Cataloguing in Publication data

Frelich, Lee E., 1957–
Forest dynamics and disturbance regimes: studies from temperate evergreen–deciduous
forests / Lee E. Frelich.
    p.   cm. – (Cambridge Studies in Ecology)
Includes bibliographical references (p.).
ISBN 0 521 65082 8
1. Forest ecology – Lake States Region.   2. Forest dynamics – Lake States Region.   I.
Title.   II. Series.
QK938.F6F74   2002
577.3′0977–dc21   2001035651

ISBN 978-0-521-65082-3 hardback
ISBN 978-0-521-05247-4 paperback

Contents

Preface

In a nutshell, this book covers the natural and settlement history of the forests in the deciduous-to-boreal forest transition zone of the Lake States (Minnesota, Wisconsin and Michigan) of eastern North America, the different types of disturbances that occur there, and how to study disturbances at the stand and landscape scales. Then several case studies from the Great Lakes Region are used to develop important concepts about the dynamic interactions between disturbance and forest size structure and composition. The dynamics of different forest types within this region are compared with each other. Finally, principles on forest response to disturbance are developed that may be generalized to temperate forests around the world. These include the dynamics of conifer–hardwood mosaics, sensitivity of stands and landscape to changing disturbance regimes, and stability at different scales.

Chapter 1 describes the forest setting of the Lake States, and Chapter 2 follows that with basic information on disturbance regimes. Chapter 3 summarizes my experiences on how to sample and analyze stand disturbance history. The techniques presented there should be applicable in most of the world's closed-canopy temperate forests. Chapter 4 summarizes what we know about stand development and successional trajectories in response to disturbance. Chapter 5 jumps to the landscape scale, and shows how to study landscape age structure and composition. Chapter 6 looks at case studies of landscape dynamics in response to complex disturbance regimes and the sensitivity of the landscape to changes in disturbance regimes, a subject which also could be called succession at the landscape scale. Chapter 7 examines how human fragmentation of the landscape changes disturbance regimes and their effects on the forest. Chapter 8 ties all of the information in the book together by examining how disturbances and biotic properties of tree species interact to structure the forest at all spatial scales from the neighborhood or small grove to the stand to the landscape. It demonstrates that four categories of landscape dynamics exist in the Lake States study area, and also that many types of forests around the world fit these same categories.

Although many excellent books have been written on individual disturbance types (e.g. Johnson 1992, Whelan 1995 and Agee 1993 on fire), few books published to date have done much synthesis of the combined effects of different types of disturbance. Therefore, I attempt to integrate the effects of disturbance regimes that are complex, with more than one type of disturbance operating at the same time. Interactions between fire and wind, and between physical disturbances and herbivory, are two interactions that are not covered very well by existing books. I make an attempt to synthesize what we know about these interactions here, and where empirical evidence is not adequate, I have not shied away from using hypothetical examples, conceptual models, and extensions of theory. In many cases, hypothetical examples can do a good job of synthesizing a complex process that was originally described in small pieces, and I employ that technique in a few places.

Acknowledgments

Although I never met him, John T. Curtis first sparked my interest in forest ecology through his 1959 book, now an ecological classic, '*The Vegetation of Wisconsin*.' It was this book that convinced me to become an ecologist. My Ph.D. advisor at the University of Wisconsin-Madison, Craig G. Lorimer, launched my studies into forest disturbance 18 years ago. Post-doctoral advisor James G. Bockheim at the University of Wisconsin introduced me to ecosystem dynamics in response to human changes in the environment. During a four-year post-doctoral with Margaret B. Davis at the University of Minnesota, I learned about long-term changes in stand and landscape development brought about by climatic change. My current collaborator, Peter B. Reich at the University of Minnesota, has generously included me in many of his research programs having to do with dynamics of the near-boreal and white pine forests of northern Minnesota. Bud Heinselman made his lifetime accumulation of knowledge on near-boreal forests available to me before his death in the early 1990s. Edward A. Johnson at the University of Calgary has been a friend throughout my career, while also unknowingly providing the inspiration to write this book by providing an example of his own Cambridge Studies in Ecology book, *Fire and Vegetation Dynamics: studies from the North American boreal forest*. Many others have helped over the years by challenging my ideas about how the forest works, including Shinya Sugita, Lisa Graumlich, Chris Peterson, and Jacek Oleksyn. Steve Friedman was very helpful in drawing the maps in the book.

Lee E. Frelich
Minneapolis, Minnesota

1 · *The forest setting*

Introduction: disturbance in temperate conifer–hardwood forests

More than one-fourth of the world's forest land lies within the cool-to-cold temperate zones of the northern and southern hemispheres. Their distinctive mosaics of evergreen conifers and deciduous hardwood species have been shaped by fire, wind and herbivory over thousands of years. In the last few centuries human activities have increasingly changed the dynamics of these mosaics. Over much of the conifer–hardwood forest zone fire frequencies have been reduced by fire suppression and exclusion, harvesting has replaced fire as the main disturbance, global warming may be causing an increase in the frequency of high winds, and the intensity of grazing has increased.

Scientists and forest managers would like to understand how changing disturbance regimes and interactions among disturbances will influence forest successional trajectories. Managers of nature reserves would like to know what types of manipulations would restore the forest to a natural condition. The main purpose of this book is to illuminate the role of disturbances in temperate conifer–hardwood forests for these scientists and managers. Therefore, I have chosen three major themes for the book:

1. To show how three major disturbance types – fire, wind and herbivory – work in combination to influence the successional trajectories and structural characteristics of forests.
2. To show how deciduous and evergreen tree species interact to form various mixtures by differentially influencing their environment and the disturbance regime. For this book, the deciduous and evergreen groups will be referred to as 'hardwoods', principally a mixture of maple (*Acer*), oak (*Quercus*), ash (*Fraxinus*), basswood (*Tilia*) and birch (*Betula*) species, and 'conifers', principally a mixture of pines (*Pinus*),

spruces (*Picea*), cedar (*Thuja*), fir (*Abies*) and hemlock (*Tsuga*) species. The common and scientific names of species referred to frequently in the book are listed in Appendix I.

3. To show how disturbance effects play themselves out over time at different spatial scales, which for purposes of discussion will be referred to as neighborhood (a small grove of trees 10–20 m across), stand (1–100 ha) and landscape (a collection of stands, >1000 ha) scales.

Forests of the Lake States region

These three themes are explored via case studies from forests in the Lake States (Minnesota, Wisconsin and Michigan, USA), which are described in the remainder of this chapter. The reader may ask why the relatively unknown Great Lakes region of the world warrants a book on forest dynamics. There are three major reasons. First is the exceptional diversity of forest types and their comparability to other forests around the world's cool-to-cold temperate zones. Second, the Great Lakes Region was settled by Europeans relatively late so that the first round of land-clearing did not occur until 1880–1940. Some large areas (14 000 to 150 000 ha), representing all of the important forest types, were protected from logging. These were influenced, but not cleared, by native Americans. Natural forces of wind and fire have been the main influences over the past several thousand years. Now that ecosystem management of forests is high priority and mimicking of natural disturbance is often incorporated in ecosystem management, the natural patterns found in the remnant areas are very relevant, and in fact desired by many forest managers. The final reason for writing a book on the Great Lakes Region is the availability of a vast scientific literature. The long-standing presence of several major universities with forest ecologists and the United States Forest Service's North Central Forest Experiment Station, along with its branches in Michigan, Wisconsin and Minnesota, means that much information is available on forest dynamics. This information has been widely scattered in many journals and research reports but it has never been presented to the scientific community in a synthesized fashion, as I attempt to do here.

The forest at the time of European settlement and today

Europeans first explored the Lake States during the 1600s. However, major settlement by large numbers of Europeans accompanied by widespread land-clearing did not occur until the mid-1800s in the southern

Table 1.1. *Forest area (thousands of hectares) in the Lake States just before European settlement (1850) and as of 1995 (Frelich 1995)*

Forest type	Forest area in 1850	Forest area in 1995	Area of primary remnants as of 1995
Jack pine	1352.9	803.9	40.7
Red and white pine	3953.9	831.0	23.1
Spruce–fir–birch	3155.4	6955.5	83.4
Swamp conifer	4272.4	1961.7	188.6
Oak–hickory	2786.7	2426.3	0.9
Riverbottom	1846.2	1605.9	3.1
Hardwood	15250.1	4670.8	29.3
Total	32617.6	19255.1	369.1

part of the region and the late 1880s to the early 1900s in the northern part. Therefore, the main questions to be answered here are: (1) How much forest existed prior to European settlement (say the mid-to-late 1800s)? (2) How much forest exists as of the 1990s? and (3) How has settlement changed the composition of the forest?

Extent and composition of forests
The distribution of major vegetation types corresponds to mean boundary positions of major air masses. The boreal forest exists north of the mean position of the arctic front during winter and during the month of June (Bryson 1966). The mixed deciduous–conifer forest exists between the boreal forest and prairie–forest border, where the arctic front sits during March and April. Thus, the prairie has long summers, the mixed forest short summers, and the boreal forest very short summers.

The Lake States included nearly 32.6 million ha of closed-canopy forests at the time of the United States General Land Office Survey just prior to European settlement, during the late nineteenth century (Frelich 1995, Table 1.1). Hardwoods, including oak–maple and maple–hemlock forests, were by far the largest component of presettlement forest landscapes, with over 15.3 million ha (47.1%), while red and white pine forest lands only occupied about 3.9 million ha, or 12% of the forest landscape (Figure 1.1). There were major differences in forest-type distribution among the three states. Nearly all of the jack pine and spruce–fir–birch forests occurred in northern Minnesota, on the Canadian Shield that has markedly colder winters and drier summers than the northern parts of Wisconsin and Michigan. The physiographic setting of northern Minnesota also allowed the development of large

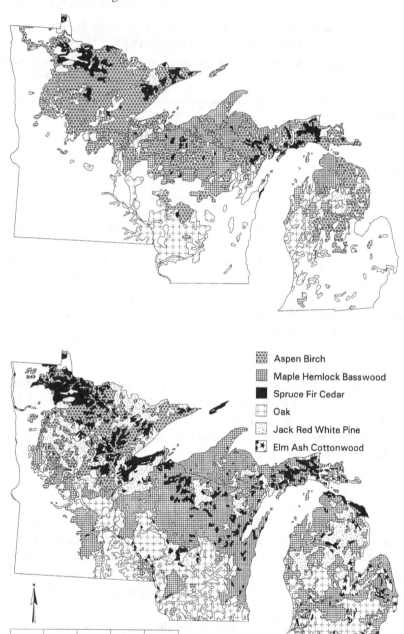

Figure 1.1 Lower panel, presettlement (*c.* 1850) and upper panel, post-settlement (*c.* 1980) forest vegetation of the Lake States. After Stearns and Gutenspergen (1987a,b).

areas of peatlands, with their associated swamp conifer forests. Michigan had the largest area of oak–hickory forest (1.5 million ha), but if oak savannas were included, both Minnesota and Wisconsin would have had twice the area of oak as Michigan (Curtis 1959, Marschner 1975). Both Wisconsin and Michigan had large areas of hardwoods, whereas Minnesota had a relatively small area of hardwoods that occurred as islands scattered within the northeastern two-thirds of the state (Marschner 1975).

The presettlement forest data can be interpreted as a stable baseline for comparison of changes in the landscape caused by humans. This is based on the knowledge that the ranges of major trees, such as maples, pines and oaks, only changed by 4–10 km/century over the last 10 000 years, and have changed little in the last few thousand years (Davis 1981). In addition, the overall rate of change in the spectrum of pollen types, on a per century basis during the 8000-year period ending prior to European settlement, was less than half that of the most recent century (Jacobson and Grimm 1986). Both of these statistics indicate great stability in area and species composition of forest in the Lake States prior to European settlement. According to United States Forest Service inventory data, there are currently 19.3 million ha forested lands – about 60% of the original 32.6 million ha (Table 1.1, Figure 1.1).

Frelich (1995) found evidence that approximately 369 000 ha of primary forest (or forest that was never logged) currently exist in the Lake States (Table 1.1). About 40% of the primary forest is in northern Minnesota's large wilderness reserve, the Boundary Waters Canoe Area Wilderness (BWCAW), and 50% is in northern Minnesota's swamp conifers. The remaining 36 000 ha is distributed among other smaller remnants, mostly hemlock–hardwood forest in Upper Michigan, including the Porcupine Mountains Wilderness State Park and Sylvania Wilderness Area, but also including substantial red and white pine at Itasca State Park, Minnesota (see Figure 1.2 for locations). The total current primary forest is about 1.1% of the presettlement primary forest of the Lake States. Percentages of original forest range from 0.02% for oak–hickory to 4.4% for swamp conifers. In addition to oak–hickory, other forest types with notably low percentages are areas of primary red and white pine (0.6%), riverbottom (0.2%) and hemlock–hardwood (0.2%) forest lands (Table 1.1).

Currently, aspen and mixed conifer–aspen stands (Figures 1.3, 1.4) occupy a much larger proportion of the forest landscape than they did prior to settlement (Figure 1.1, Table 1.1). This is due to extensive forest

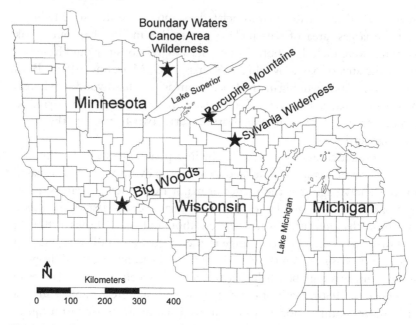

Figure 1.2. Location map of the Lake States Region, showing the major study areas.

Figure 1.3. Young quaking aspen stand typical of second-growth forest in the Lake States. Photo: University of Minnesota Agricultural Experiment Station, Dave Hansen.

Figure 1.4. Mixed aspen, white spruce and balsam fir forest was/is common in the northern Lake States region prior to European settlement and today. Photo: University of Minnesota Agricultural Experiment Station, Dave Hansen.

clearing followed by burning of slash that occurred between 1850 and 1940 in the Lake States. In northern parts of Minnesota and Michigan, forests of spruce and jack, red or white pine (Figures 1.5, 1.6, 1.7) yielded to aspen, while in northern Wisconsin and parts of northern Michigan, hemlock–hardwood or hardwood forest (Figure 1.8) was converted to aspen. Lowland conifer forests (Figure 1.9) have seen relatively little conversion to aspen or other forest types, due to the undesirable sites they occupy.

Climate

The climate of the Great Lakes Region is humid continental. Summers are short and cool; average July temperatures range from 17 °C in northern Minnesota to 19–20 °C in the Upper Michigan study areas, to 22 °C in the oak–maple forests of southern Minnesota. Winters are long and cold; average January temperatures range from −17 °C in northern Minnesota to −6 to −8 °C in Upper Michigan and southern Minnesota (National Oceanic and Atmospheric Administration 1980). It should be noted that there is a strong lake effect and that temperatures within

Figure 1.5. The southernmost occurrence of upland black spruce forest occurs in Minnesota's Boundary Waters Canoe Area Wilderness. Photo: University of Minnesota Agricultural Experiment Station, Dave Hansen.

Figure 1.6. Even-aged jack pine forests occur after fire on sandy and rocky sites along the southern margin of the boreal forest. Photo: University of Minnesota Agricultural Experiment Station, Dave Hansen.

Figure 1.7. An old-growth red pine stand at Itasca State Park, Minnesota. Photo: University of Minnesota Agricultural Experiment Station, Dave Hansen.

Figure 1.8. Sugar maple stand typical of the 'Big Woods' region of Minnesota and the northern mesic forests of Wisconsin and Michigan. Photo: University of Minnesota Agricultural Experiment Station, Dave Hansen.

Figure 1.9. Lowland black spruce on sphagnum peatlands are extensive in northern Minnesota, Wisconsin and Michigan. Photo: University of Minnesota Agricultural Experiment Station, Dave Hansen.

10 km of the Great Lakes may be higher during the winter and lower during the summer than indicated above (Eichenlaub 1979). Continentality (an index of annual temperature range) in Upper Michigan is nearly the same as the New England coast of Maine (Trewartha 1961). Mean annual frost-free period ranges from about 90 days in the northern part of the region (although about 120 days near Lake Superior), to 160 days in the southern part (Phillips and McCulloch 1972). Day lengths range from about 8–9 hours on December 22 to 15–16 hours on June 21.

Annual precipitation ranges from 800 mm to 900 mm over the region and is much higher during the summer months than winter months, except near Lake Superior, where it is fairly evenly distributed throughout the year (Eichenlaub 1979). Measurable precipitation (0.25 mm or more) falls on 130 to 160 days per year near the Great Lakes, which is the same range as maritime areas in the Pacific Northwest and New England, USA (Court 1974). Away from the Great Lakes, most precipitation falls during intense convective storms during summer months, so that measurable precipitation falls on only 100 days. Only 5–10% of months in Upper Michigan can be considered to have severe or extreme drought (Court 1974). About 25–35 thunderstorm days occur annually in the region (Court 1974), but owing to accompanying rain, these result in only 1–5 lightning fires per 400 000 ha per year (Schroeder and Buck 1970). Annual snowfall is 1–2 m away from Lake Superior, but in Lake-effect snowbelts, average snowfall may be as high as 4 m, and occasionally up to 7 m of snow falls in one winter in the Porcupine Mountains (Muller 1966, Eichenlaub 1979). The ground is snow covered on and off from December to March in the southern part of the region, and continuously from late November until mid April in the northern part. The region receives about 60% of possible sunshine during the summer months and only 30–40% of possible sunshine during the winter months (Visher 1954).

The principal primary forest remnants

Hemlock–hardwood forests of Upper Michigan

In Upper Michigan there are two large areas of forest that have escaped logging. The largest is the Porcupine Mountains Wilderness State Park, on the coast of Lake Superior near the western end of Upper Michigan (Figure 1.2). Braun (1950) presents a description of the vegetation of the Porcupine Mountains and calls the area 'a vast hardwood forest containing the most extensive primeval tracts remaining on the continent.' It should be noted, however, that some of the area has (since 1950) been logged. Even though acquisition of the area was authorized in 1944, not all of the property was immediately purchased. Large-scale logging operations, which did not reach western Upper Michigan until the 1940s, were under way as the park was being purchased. Today, the park contains approximately 14 500 ha of forest which was never logged, although about 1500 ha had some fallen timber salvaged after a windstorm struck the area in 1953. The second-largest area (6073 ha) of primary forest is in the Sylvania Wilderness Area in Ottawa National Forest (Figure 1.2). Sylvania is a township on the Wisconsin–Michigan border 80 km from Lake Superior which was held as a private preserve until being sold to the United States Forest Service in 1966. The only trees removed were in a forest thinning near roads and a few pines on the lakeshores.

The principal advantage of conducting studies of forest dynamics on a few large blocks of remnant forest (as opposed to many small ones) is that a bias toward old-growth stands with large trees is avoided. Many of the small remnant forests, such as smaller state parks, scientific areas and Nature Conservancy preserves, were selected for purchase specifically because they contained old-growth forest. Young forests developing after natural catastrophes were probably not considered for purchase because the forest was not recognized as being much different from the ubiquitous second-growth. In general, these small areas do not contain young virgin sapling and pole-sized stands as do the large blocks of forest which were preserved for other reasons.

The principal vegetation type on all these forest remnants is northern mesic forest as defined by Curtis (1959). Sugar maple dominates most of the forest inland from Lake Superior, and mixes extensively with hemlock near the lake. Occasional stands of hemlock are also found inland, especially in Sylvania. Lesser amounts of yellow birch, red maple, basswood and white pine occur throughout the area (Braun 1950). All other tree species are of local or sporadic occurrence. Elevations in the Porcupine Mountains range from 182 m on the surface of Lake Superior to about 600 m at 5 km inland. Glacial Lake Duluth covered parts of the area until approximately 8000 ybp (Hough 1958). These lake–plain areas have very deep deposits of silty lake-bottom sediments at elevations up to 120 m above Lake Superior, with deep loam and silt loam soils predominating on gentle north slopes (0–10%). Farther inland, bedrock comes near the surface and topography is more rugged, with slopes up to 30% in steepness on the predominating north slopes. Soils are still generally 1 m or more deep in these upland areas and are of loam or sandy loam texture. South slopes in both areas have very steep or vertical outcrops of bedrock unsuitable for development of forests, although a few gently sloping south-facing hills are present. Soils include coarse-loamy, mixed, frigid Alfic Fragiorthods and Entic Haplorthods (Michigan State University 1981). Observations of thickness of root mats of recently fallen trees indicate that fragipans occur at depths of 50–100 cm. In the uplands, Lithic Haplorthods and Lithic Borofolists also occur locally.

The topography of Sylvania is quite different from the Porcupine Mountains. A glacial end moraine known as the Watersmeet Moraine covers the area (Dorr and Eschman 1970, Jordan 1973). The pitted ice-contact topography varies from 500 m to 550 m in elevation and contains many small lakes and bogs. The drift is deep (>30 m), red color, slightly acidic, and of sandy loam texture. The upland forest soils at Sylvania are spodosols and most are classified as coarse-loamy, mixed, frigid Alfic Fragiorthods or Alfic Haplorthods (Jordan 1973).

Near-boreal forests of northeastern Minnesota
A small extension of boreal forest extends from Canada into northeastern Minnesota (Weber and Stocks 1998). Because it is on the very southern

edge of the boreal forest and contains a few scattered stands of species not usually considered boreal, Heinselman (1973) coined the term 'near-boreal forest,' which I adopt here. The boreal tree element includes jack pine, black spruce, and aspen, mixed with balsam fir, white cedar, white spruce, and paper birch in older stands. The non-boreal element exists in stunted form here at the northern edge of the range, including white and red pine, red maple, northern red oak, bur oak, and pin oak. Red maple and the oaks in particular can be in the form of a shrub or small tree, unlike the large sizes (1 m diameter at breast height, dbh) they reach further south. Near-boreal forests include the following types: (1) fir–birch forest on relatively good soils; (2) jack pine–black spruce on coarse shallow soils over granitic bedrock, as well as several other jack pine-dominated types; (3) red maple, aspen, birch and fir on moist but not wet sites; (4) red pine on shallow rocky soils, especially common along lakeshores; and (5) birch–white pine forests, common along lakes and streams regardless of the soil type. A variety of conifer swamp forests with black spruce, tamarack or a mixture of the two also occurs.

The Boundary Waters Canoe Area Wilderness lies within the near-boreal forest zone (Figures 1.1, 1.2), centered at about 48 °N latitude, and 91 °W longitude. It contains about 439 000 ha of land and water and was set aside as a wilderness area off limits to logging by the United States Congress in 1978. Of this, 335 000 ha is forested, and 169 000 ha (50% of forest) has never been logged (Heinselman 1996). The stand-origin dates for the entire 160 000 ha that was never logged were mapped by Heinselman (1973, 1996), and the community makeup of this tract was also sampled by Ohmann and Ream (1971). Fourteen major fire years, with large stand-killing fires, have occurred since 1595, including four (1863–64, 1875, 1894, and 1910) that account for nearly three-quarters of the forest area.

'Bigwoods' forests of southeastern Minnesota
These forests were at the edge of the prairie–forest border in the south-western part of the Lake States (Figure 1.2). Early survey data indicated a large patch about 8000 km² in size that was dominated by a mixture of red oak, white oak, American elm and sugar maple (Grimm 1984). This patch was typical of forests along the prairie border in the Mid-western United States, in that the oaks were more abundant along the prairie edge, and the maple and elm more abundant in the interior of the forest, and the forested area was often on the northeast side of some sort of topographic feature that served as a fire break (Grimm 1984, Leitner *et al.*

1991). Soils were relatively deep loams to sandy loams on top of glacial drift.

At this point, only a few small scattered remnants of Bigwoods vegetation exist, which may comprise 1% of the original forest (Minnesota Biological Survey 1995). These remnants are tiny compared with the Porcupine Mountains or Boundary Waters Canoe Area Wilderness. As they are the only representatives of a formerly large forest, however, their importance goes beyond their size. Because of fire suppression during the past century, sugar maple has increased in abundance, while oaks have decreased in many stands, such as that in Figure 1.8.

Summary

The Lake States are in a unique zone of sharp climate change so that a large variety of forest types is present. A large body of scientific information exists on the response of these forests to disturbances. In this book, I synthesize these findings so that Lake States forest dynamics can be compared with analogous forests in the world's cool-to-cold temperate zones. Lake States forests include oak and beech–maple forests analogous to those in western Europe, Japan, southern South America and New Zealand; hemlock forests similar to those in eastern Asia; red and white pine forests with analogs in northern Korea, China, and northern Europe; vast peatland larch and spruce forests, analogous to those in Siberia; and fire-adapted near-boreal forests of jack pine and black spruce, similar to those across southern Canada, northern Scandinavia, and northern Japan. Several large primary forests still exist in the Lake States so that the dynamics of forests with minimal human influence can be compared with other regions where human influence has had more impact. Enough detail on the Lake States forest history, composition and dynamics is presented to facilitate comparison with other forests. I hope to enable readers to qualitatively predict forest response to disturbances in any of these other temperate conifer–hardwood forests around the world.

2 · The disturbance regime and its components

Importance of disturbance in forests

Disturbances exert strong control over the species composition and structure of forests. As a general rule, landscapes with frequent severe disturbance are dominated by young even-aged stands of shade-intolerant species such as aspen. Conversely, old stands of shade-tolerant species such as hemlock dominate where severe disturbances are rare. Every conceivable mixture between these two extremes can be created by the various combinations of disturbance. To understand how disturbances exert these influences over the forest it is necessary to know the basic concepts and mechanics of disturbance – the function of this chapter. Fire, wind and herbivory have been chosen for detailed discussion because they are very important influences on temperate forests and we know a lot about them and their interactions.

A definition and key concepts

The disturbance regime is simply a description of the characteristic types of disturbance on a given forest landscape; the frequency, severity, and size distribution of these characteristic disturbance types; and the interactions among disturbance types. If the forest experiences a series of unique disturbances over time, so that type, frequency, severity and size cannot be characterized, then there is no stable regime. Apparent stability of the regime, however, is a function of the length of time and size of area observed (Lertzman and Fall 1998). A paleoecologist may view the temperate forest of eastern North America as unstable, because it started out as spruce after the last glaciation receded, changed to pine or oak as the climate reached its maximum interglacial warmth 7000–9000 ybp, and then converted to what we know today as 'hemlock–hardwood forest' 3000 ybp. The forest experienced a series of unique climatic and disturbance events that resulted in major changes. From the forester's point of view, however, there may have been periods of stable disturbance regime at time scales of decades to centuries between major changes. It is becoming increasingly obvious that forest change over time exhibits a punctuated stability phenomenon, and those who view the vegetation as stable and those who view it as unstable may both be right (see Chapter 8).

Disturbance intensity versus severity

The difference between disturbance intensity and severity is important to grasp for anyone involved with research in forest ecology or forest management. Intensity refers to the amount of energy released by the physical process of disturbance, and severity refers to the amount of mortality that occurs among tree and plant populations in a disturbed area. With fires, intensity is the rate of heat energy released per unit length of fire line per unit time (W/m/s), or sometimes the rate of heat produced per unit time and area (W/m^2/s). In addition, fire intensity has a direct relationship to flame length (Johnson 1992). Often there is a good correlation between fire intensity or flame length and severity. However, in areas with organic soil, whether bogs or rocky areas where the soil is a moss blanket, a low-intensity fire (perhaps with flame lengths less than a meter) can kill the roots and result in nearly 100% tree mortality. Such a fire would be low in intensity but high in severity. Regarding windstorms, intensity and resulting severity are highly correlated. The higher the peak wind gust (i.e. the higher the amount of force on the trees), the greater the proportion that will blow down. Thus, for windstorms intensity and severity have the same meaning in terms of the outcome for the forest, even though the units are different. Herbivory (at least in temperate forests – we do not have armies of locusts or ants) always has a very low intensity (i.e. rate of consumption), because animals or insects consume relatively small amounts of biomass over periods of days, weeks, or years. The severity, however, can range from virtually no mortality and impact on the forest, to nearly total mortality of trees at the stand or landscape level. For convenience, we can arrange disturbance severity into three ordered categories:

1. Low-severity disturbances are those that kill small pieces of the forest understory or overstory (or both), resulting in scattered minor mortality. Windstorms that pick off a few larger trees and create scattered

treefall gaps, spot fires, and selection cutting of single trees or small groups of trees are examples.

2. Moderate-severity disturbances kill most/all of either the understory or overstory, but leave a substantial legacy of intact mature trees or seedlings. Windstorms and clear cutting that remove the canopy but leave the seedling layer intact, and surface fires or browsing deer that kill nearly all seedlings while leaving the canopy intact, are examples.

3. High-severity disturbances kill most of the understory and overstory layers. Crown fires and clearcutting followed by burning of the remaining slash are examples.

Rotation period

Rotation period is defined as the length of time required to disturb an area equivalent to the whole landscape of interest. Synonyms used include fire cycle (Van Wagner 1978), which is the rotation period for stand-killing fires, and mean recurrence interval. Rotation period may be calculated for any type of disturbance (wind, surface fire, crown fire, logging, etc.), so that any one forested landscape will have several 'rotation periods' in effect. 'Natural rotation period' refers to the rotation for some sort of natural disturbance, while 'commercial rotation period' refers to the ideal logging rotation for lumber or fiber production. The principal difference between the natural and commercial rotation period, in a functional sense, is that stands to be disturbed are 'selected' much more randomly under the natural disturbance regime, and more purposefully under the harvesting regime. The implications of this difference and methods for estimating rotation periods are covered in Chapters 5 and 6.

The phenomenon of windstorms

The Great Lakes Region (as well as cold-temperate forest zones around the world) is one of the most active weather zones in the northern hemisphere, with the polar jet stream positioned overhead much of the year (Bryson 1966, Eichenlaub 1979). More cyclones pass over the area than any other part of the continental United States (Visher 1954, Whittaker and Horn 1982). Major cyclone tracks cross the region every month of the year (Klein 1957, Whittaker and Horn 1982). Outbreaks of severe weather are known to occur in the region when the polar jet stream lies just to the north and the subtropical jet stream lies just to the south

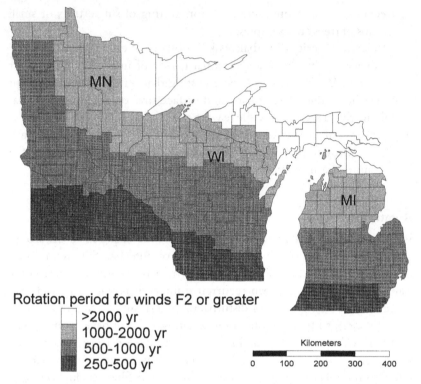

Figure 2.1. Approximate rotation period for winds F2 or higher (>180 km/h) during the historical period 1850–1970. Assumes that straight-line thunderstorm winds affect twice as much area as tornadoes, based on author's stand-history reconstructions and tornado data from Thom (1963).

during the summer months (Eagleman *et al.* 1975, Whitney 1977, Doswell 1980). Some of the most severe thunderstorms ever observed by meteorologists have occured in northern Wisconsin and Minnesota (Fujita 1978). There is a gradient in the frequency of occurrence for all types of severe windstorms in the Lake States (Figure 2.1), running from the southwest (maximum occurrence) to the northeast (minimum occurrence).

Topography affects forest damage caused by wind. Wind speeds are usually strongest at the top of a slope if wind is at a right angle to a ridge, at the midslope if wind is at an acute angle to a ridge, and at the valley bottom if wind direction is parallel to a ridge. Turbulence caused by eddies as the wind swirls around topographical obstacles also can cause

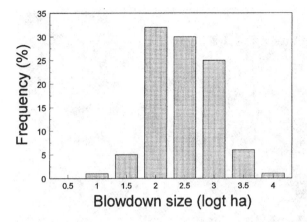

Figure 2.2. Size distribution of historic forest blowdowns in the Lake States (after Fujita 1978 and Canham and Loucks 1984). Logt stands for log base 10.

significant tree damage, particularly roller eddies on the lee side of ridges (United States Department of Agriculture 1964, Gloyne 1968). There is usually a strong relationship between forest damage and aspect in areas hit by tropical cyclones, where damaging winds may frequently come from the same direction (Boose *et al.* 1994). However, the overall pattern of damage may not correspond to topographical features in areas where winds come from all directions and/or the winds have a downward component, as in derechos, downbursts and tornadoes (Rebertus *et al.* 1997, Frelich and Lorimer 1991a). The damage pattern from one storm may have a directional component, but as numerous storms with different wind directions overlap, the correspondence with aspect is washed out. Individual blowdowns in the Lake States ranged from from 0.1 ha to 3600 ha, with a mean of 93 ha during the presettlement era (Canham and Loucks 1984, Figure 2.2). There are three major storm types that cause forest gaps, discussed below.

Straight-line thunderstorm winds

Downbursts and microbursts are the forms of straight-line winds from severe thunderstorms that cause most forest damage. Severe thunderstorms with downbursts or tornadoes can be identified visually as they approach by the mammatus clouds on the underside of the storm's anvil cloud (Figure 2.3). The only difference between downbursts and

Figure 2.3. A severe thunderstorm (indicated by mammatus clouds on the underside of the anvil, pictured here) rolls across southern Minnesota during the summer of 1994. Such storms are capable of producing downbursts, derechos, and tornadoes. Photo: University of Minnesota Agricultural Experiment Station, Dave Hansen.

microbursts is the size, with those having a damage swath more than 3 km in length referred to as downbursts (Fujita 1978). Both consist of a shaft of cold dense air that accelerates downward, hits the ground, and splatters outwards in all directions. Because thunderstorms usually are moving and each downburst takes several minutes to form, the resulting damage fields are oval in shape and elongated in the direction of storm movement. The two components of the wind are: (1) the forward momentum of the thunderstorm or squall line, commonly in the range of 50–80 km/h; and (2) energy from the jet stream which may interact with the storm to accelerate the downward wind movement. Wind speeds in the most severe downbursts are F2 on the Fujita scale (180–250 km/h) and result in complete forest canopy blowdown.

One thunderstorm may produce many downbursts over a period of a few hours, and the resulting pattern of damage is known as a downburst family, also called a 'derecho' (pronounced der-ay'-sho). Derechos can affect areas 100 km or more in width, and may travel across half of the

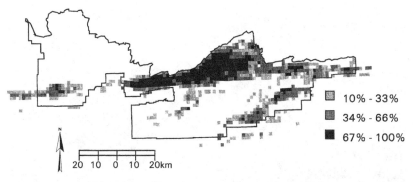

Figure 2.4. Superblowdown of northern Minnesota, July 4, 1999. Several such blowdowns in recent years suggest that the historic size distribution of forest blowdowns has changed during the last century.

North American continent. On May 30–31, 1998, a derecho formed near Minneapolis, Minnesota, and traveled all the way to the Atlantic Ocean, 2000 km to the east, causing significant tree damage all along the way. Areas of heavy canopy blowdown, however, are typically restricted to areas about 10–25 km in width and 100–200 km in length.

Three infamous derechos have caused massive damage to Lake States forests in recent decades, across northern Wisconsin (1977), north of Itasca State Park in Minnesota (1995), and the Boundary Waters Canoe Area Wilderness in northern Minnesota (1999). The Wisconsin storm, called the Flambeau Blowdown, was a family of 26 separate downbursts that cut a swath of damage 20 km wide and 200 km long across northern Wisconsin on the 4th of July, 1977 (Fujita 1978). Wind speeds reached 250 km/h in some areas and 140 000 ha of forest had at least 50% of the canopy removed (Fujita 1978, Frelich and Lorimer 1991a). The Minnesota storm, (which also occurred on the 4th of July!), caused significant damage to 200 000 hectares of forest in northern Minnesota's Boundary Waters Canoe Area Wilderness (Figures 2.4, 2.5). These recent massive forest blowdowns are an order of magnitude larger than those observed in the historic data set (Figure 2.2), and this raises the question of whether climate change has begun to increase the size and severity of thunderstorms in the Lake States. An alternative explanation is that the size distribution of blowdowns has not changed over the last century and that by chance none of these large blowdowns happened to be present and recorded at the time of the presettlement land survey. Further observation and analyses will be necessary to test these hypotheses.

Figure 2.5. A 198-year-old red pine forest lies in shambles after the superblowdown of July 4, 1999, northern Minnesota. Photo: University of Minnesota Agricultural Experiment Station, Dave Hansen.

Tornadoes

These vortices of rapidly rotating air are produced by rotating supercell thunderstorms, and they typically have damage paths 20–30 times, rarely up to 100 times, as long as wide (Abbey and Fujita 1983). The maximum width is 2 km, but a typical width is 100–200 m. Tornadoes can range from F0 to F5, with those F2 and above capable of causing extensive forest canopy damage. Although only about 3% of all tornadoes reach F4 or more in intensity (with wind speeds of >320 km/h), they cause the majority of all damage, because they have the widest and longest paths (Fujita 1978, Abbey and Fujita 1983). Forest damage from a tornado is characterized by trees falling in convergent directions, as opposed to straight-line thunderstorm winds, which are divergent with a much shorter length-to-width ratio than tornadoes.

Tornadoes occur in all of the earth's temperate zones. However, the probability of being hit by a tornado, as well as the F-level of tornadoes, reaches its maximum in the interior of North America, in a region known as 'Tornado Alley', stretching from Texas and Oklahoma, through Kansas, Missouri, Iowa and Illinois in the central USA (Thom 1963, Eagleman et al. 1975). The Lake States study area of this book is

Figure 2.6. A tornado churns through the Midwestern USA. Photo: U.S. Department of Agriculture.

just off the north end of 'Tornado Alley', and experiences very intense tornadoes (Figure 2.6). However, the annual probability of being hit by a tornado ranges from 5% to 40% of that in central Oklahoma, where the highest frequency occurs (Thom 1963). These values correspond to tornado rotation periods of about 6000 years in the extreme northern part of the Lake States to about 1000 years in the southern part.

Cyclonic winds (gales)

These are general winds around low pressure centers, also called extra-tropical cyclones. Very large areas are covered in individual storms (millions of km^2), but the intensity of the wind is lower than tornadoes and straight-line winds. In the Lake States, extra-tropical cyclone wind speeds rarely reach 120 km/h for several hours, causing low-to-moderate forest canopy damage in the Lake States. However, extra-tropical cyclones can be more severe near the ocean, in areas such as Great Britain and southern South America, where stand-leveling winds of 180 km/h can occur (Gloyne 1968, Rebertus *et al.* 1997). Tropical cyclones have similar or higher wind speeds up to 280 km/h, and occasionally affect the temperate zone forests in late summer or early autumn, in the New

England region of the USA, and in Japan (Boose *et al.* 1994, Ishizuka *et al.* 1998). These severe extra-tropical and tropical cyclones don't affect the Lake States. Nevertheless, given the Lake State's winds of 250 km/h and >300 km/h during severe thunderstorms and tornadoes, the severity of the wind regime is probably as high or higher than that in regions near the oceans.

Accumulations of heavy wet snow or freezing rain accompanied by cyclonic winds of 50–100 kph may also occur in November and April in the Great Lakes Region, and these may cause loss of a small percentage of canopy trees over a large region. There may be a few areas 1–10 ha in size where most of the canopy is taken down in such a storm, especially in mature near-boreal forests that typically have tall, thin jack pine or black spruce trees. Crown damage, rather than trunk breakage or uprooting, is common in white pine from these storms.

There is a relationship between cyclones, thunderstorms and tornadoes. Many intense cyclones have gale-force general winds from the northeast, southwest and northwest, in response to the atmosphere's attempt to fill the low pressure area, while a cold front trails along from near the center to the southwest. These cold fronts, and/or the dry line in front of them, are where severe thunderstorms form that may produce microbursts, downbursts, derechos and tornadoes. Especially in the spring and fall, it is common for an intense northeast or northwest gale to be blowing in northern Minnesota, Michigan and Wisconsin, while severe thunderstorms are in progress in southern Minnesota and Wisconsin.

The phenomenon of fire

On the presettlement landscape there was large variability in the severity of fires and their frequency across the forests of the Great Lakes Region. Near the prairie–forest border, fires were very frequent, having occurred once or twice a decade. These areas were covered with savanna-like vegetation with scattered groves of trees embedded within grasslands. The general trend as one moved toward the northeast was for fire frequency to decrease but for fire severity to increase. Thus, the near-boreal forest mixture of conifers and birch and aspen that covered much of northern Minnesota had very severe canopy-killing fires with average fire rotation periods ranging from 50 to 200 years among forest types. Fire was generally rare in hemlock–hardwood dominated landscapes.

The United States Forest Service has designated the fire climate of the Great Lakes as Region 11, which has an April to October fire season. In hardwood forests fire frequency peaks before leafout in spring and after the leaves have fallen in autumn, when wind and sunlight are able to dry the forest floor (Schroeder and Buck 1970). Fuels in the hardwood forests consist mainly of compact litter and branchwood that will support surface fires. Crown fires are very rare in hardwood forests due to relatively (compared with many conifers) high foliar moisture content, low bulk density of the canopy and possibly low content of flammable extractives (Van Wagner 1977). The reader should consult Johnson (1992), Agee (1993) and Whelan (1995) for more detail on the physics and effects of fires.

Several very large and intense fires, such as the Peshtigo Fire which burned about 500000 ha during October 1871, have occurred in the Lake States region, even in areas where the main vegetation type is the non-flammable hardwood forest type. Two conditions were necessary for the development of these conflagrations. First, logging and burning to clear land for settlement was extensive at the time of the fires and heavy accumulations of mixed fuels (logging slash) were widespread (Wells 1968). Second, these conflagrations were preceded by 3–5 months of below normal rainfall and several days of low humidity (Haines and Sando 1969). Such conflagrations are unlikely to occur again (even though the meteorological conditions may be repeated) because modern forestry practices do not create such vast areas of logging slash on the landscape. However, it should be kept in mind that conflagrations may have occurred in the presettlement forest in areas where extensive blow-downs resulted in fuel conditions similar to logging slash.

Most northern conifer forests have canopy foliage with high bulk density, closed canopies and heavy ground fuel loadings. Under these conditions intense crown fires with flame lengths of 20–30 m are possible, especially after long droughts. The actual fire-line intensity in these forests is much more controlled by the fire weather at the time of fire than by any other factor (Johnson and Fryer 1989, Johnson and Wowchuk 1993, Bessie and Johnson 1995). If there has been a severe drought and the day of a fire includes low humidity and high wind speeds, severe crown fires beyond human control capabilities are likely, whereas a moderate drought will likely lead to surface fires or passive crown fires. Fires with flame lengths of more than 5–10 m may be uncontrollable. As a general rule, people can start and stop surface fires, but they can only start crown fires. Most crown fires are finally put out by rainfall or snowfall events that bring the period of severe burning weather to an end.

Fires in northern Minnesota's vast conifer forests prior to European settlement were very large in size, with a mean of 4000 ha and a maximum estimated size of 160 000 ha (Heinselman 1973). The average fire size means little in northern conifer forests, because most of the area is burned during the top few fires in terms of area. A good rule of thumb is that the top 3% of fires in terms of area within the fire perimeter burn 97% of the landscape – note that the biggest fire cited burned 40 times the area of the average fire. Ultimate fire size is determined by the length of time with severe burning conditions, especially the number of days with low humidity and high winds that are brought to an end by precipitation, and the placement of major firebreaks, such as large lakes. Thus, the maximum fire size varies with landscape characteristics.

People have an important influence on the size, severity and frequency of fires, and these have changed in some areas since settlement by Europeans. Conversion of much of the landscape from conifers to aspen and birch (which do not carry severe crown fires very well), to cities, farms, highways and resorts, has disrupted the flow of fires across the landscape, and fire now plays a more minor role in Great Lakes forests than it used to. These changes in land use result in *fire exclusion* (often confused with *fire suppression* by many people) that could account for much of the reduction in the role that fire has played in the forest over the last century. In addition, direct suppression of fires and climate change have probably also contributed to the reduction in fire frequency. Large fires are still possible and will eventually occur in northern Minnesota – the annual probability is just lower at this point than in the past.

Johnson (1992) argues that direct suppression of fires has played a minor role in the reduction in fire frequency over the last century in North American near-boreal and boreal conifer forest. The logic for this argument is that 97% of the landscape is burned during the largest 3% of the fires. These very large fires only occur when there is a prolonged sub-continental-scale drought, and any small source of ignition blows up into a conflagration with flame lengths that are considered beyond the abilities of fire suppression. Suppression can put out the other 97% of fires, but these would have been relatively low-intensity events in any case, and would only burn 3% of the landscape even if no suppression occurred. Therefore, the logic goes, suppression changes the area burned in northern conifer forests very little. The one flaw in this argument pointed out by some is that some fires are ignited by lightning when the forest is relatively wet, and then smolder for several weeks until a severe drought develops, and then take off and become big fires (e.g. Lorimer and Gough 1988). During this smoldering phase, suppression would be effective in stamping out an ignition source before it starts a big fire.

However, Johnson counters that the empirical evidence from remote parts of the boreal forests where fire suppression is not attempted shows that fire frequency and rates of landscape burned per decade change at the same times as they do in regions where suppression is practiced (Johnson 1992, Johnson and Gutsell 1994).

The phenomenon of mammalian herbivory

Several members of the deer family, particularly white-tailed deer (*Odocoileus virginianus*) and moose (*Alces alces*) in North America and red deer *(Cervus elaphus)* in Europe disturb the forest by *browsing* of woody seedlings and saplings during the winter and *grazing* of herbaceous plants during the summer. Rodents such as mice, squirrels and voles, and rabbits also eat tree seeds and small seedlings, sometimes altering the future course of succession (Ostfeld *et al.* 1997). Browsing and grazing do not kill adult trees, and this type of very low-intensity disturbance must continue for several decades to influence future forest composition by preventing successful seedling establishment and canopy recruitment of certain tree species. One extreme example of long-term browsing of tree seedlings has occurred in Scots pine forest near Braemar, Scotland, where red deer have prevented successful recruitment of new Scots pine since the late 1700s (Watson 1983).

Ungulates do not always browse the same tree species in different regions, even when those same species are available. One reason for this is that several available browse species in each forest can be placed in an order of preference (e.g. Beals *et al.* 1960, Anderson and Loucks 1979). Deer and moose usually choose to eat those on the 'highly preferred' list, while causing only minor damage to the 'moderately preferred', or 'not preferred' lists. When the most preferred species is absent, the second most preferred species may become the main food source and suffer major damage. A second reason for variability in which species are taken is that palatability of the same tree species can vary among regions. This is probably an evolutionary adaptation by the trees to a long history of heavy browsing. The lack of palatability is caused by certain secondary compounds in the twigs.

Deer and moose in the Lake States

Deer have generally increased greatly in abundance over the last 50 years, after a period of near extirpation during the land-clearing of the late

1800s to the early 1900s. Their populations irrupted throughout the region beginning in the 1920s (Leopold 1943) in response to protection and regrowth of forest in the early-successional types that are favored by deer. Moose have generally declined, except in a few instances, such as on Isle Royale National Park in Lake Superior, where moose swam to the island in the early 1900s, and thrived in the absence of competing members of the deer family (Murie 1934).

During winter browsing, moose prefer paper birch, mountain ash and mountain maple, and to some extent balsam fir, over other species of trees. Moose simply grab a sapling and pull, stripping small twigs and leaving a ragged branch stub. They can also bend and break saplings as large as 2.5 cm in trunk diameter for access to twigs too high to reach directly. Thus, seedlings are very unlikely to escape moose browsing by growing too large. There are no tree species in the Lake States that can attain a dbh of 2.5 cm during one summer.

White-tailed deer prefer several species of woody plant for winter browsing. Severe regeneration problems have been documented for white cedar, yew and hemlock among the conifers, and red maple, red oak and white oak among the hardwood species (Beals et al. 1960, Frelich and Lorimer 1985). Damage to white pine varies throughout the region. In northern Minnesota, deer prefer white pine, but they generally do not eat much white pine if one of the highly preferred species, such as red oak, is present in a given stand. Conifers with long leaf lifespan are less able to recover than deciduous tree species after browsing. Hemlock seedlings, for example, can die after one defoliation, whereas maples and oaks can be repeatedly browsed and persist as a shrub. Vigorously growing hardwoods in forest gaps can sometimes also escape deer browsing after a delay of a few years. The mechanism for escape is that deer can only browse stems up to 1 cm in diameter. A new leader on a hardwood may be 0.5–0.8 m long, and the bottom 0.2–0.3 m of it is often beyond the diameter that deer take. Thus, the saplings in gaps advance 0.2–0.3 m per year, and escape deer browsing when they reach a height of ≥2 m. Conifers of the preferred species have thin leaders that are vulnerable to total removal by deer, and because of a conifer's excurrent growth form, it is hard for them to re-establish a new leader.

Deer overwinter in certain areas, called deer yards, which provide shelter from predators and winter winds. Significant tree damage is caused in these deer yards. During the summer, deer spread throughout the forest, grazing mainly herbaceous plants. Little summer damage is caused except in fragmented forests, where farm fields with abundant food allow maintenance of high deer populations that then use the forest for cover and may destroy understory tree and herb populations in the process. Species in the lily family (geophytes) have been locally extirpated (Augustine and Frelich 1998). They are most susceptible because they only put forth one growth spurt each summer. If the top is grazed, then the plant cannot store much energy in the bulb for next year's flush of growth. Other plants can simply regrow from the point of grazing and are not badly damaged unless the grazing is continuous and takes away all the flowers so that reproduction cannot occur.

Insect damage

Several species of insects defoliate trees, usually not leading to mortality except in trees weakened by some other stress. There is one major native insect species, spruce budworm (*Choristoneura fumiferana*), that defoliates and kills balsam fir and white spruce along the southern edge of the boreal forest (Blais 1983, McCullough *et al.* 1998). Outbreaks typically last anywhere from 1 year to 25 years, and there is a marked periodicity between peaks in defoliation of 31–37 years in the southern boreal forest of eastern Canada (Hardy *et al.* 1980, Candau *et al.* 1998). During the peak periods, spruce budworm causes almost complete canopy mortality in monospecific stands of fir. In some areas where fir is mixed with other species in a mosaic, there is chronic infestation of fir, which mainly kills larger trees and shortens their mean lifespan. However, in these mosaics, there are virtually always young fir that are not badly infested so that the fir component of the stand continues indefinitely (Heinselman 1973, Bergeron *et al.* 1995).

Disturbance interactions

The properties of the principal disturbance types have been described and an examination of interactions among disturbance types, and interactions of the vegetation with disturbance, are the next logical subjects to consider. These interactions are introduced here as background for the next several chapters. They are also summarized in conceptual model form in Chapter 8.

Disturbance–vegetation interactions

Fire, wind and tree mortality
As a general rule, high-intensity crown fires with a high rate of spread kill trees of all sizes, or at least the above-ground parts in the case of sprouters.

After passive crown fires, the proportion of the canopy scorched and the proportion of the circumference of the basal cambium killed jointly determine the probability of death, which may be immediate or within several years of the fire. With surface fires, the smaller a tree the more likely it is to be killed. The degree of insulation provided by the bark and length of time that heat is applied to the outside of the bark determine the probability of cambial death. Bark thickness is greater for larger diameter trees of all species in the Lake States. Fire-tolerant species generally attain larger diameters and have thicker bark than fire-intolerant species even when comparing trees of equal diameter, and their bark may be a better insulator for an equal bark thickness than for fire-intolerant species. Therefore, very large differences in length of burning time required to kill a tree may result. For example, white pines are often 30–40 cm dbh in a stand where the balsam fir are only 10 cm dbh, and the pines are able to sustain surface-fire burning 4 to 6 times as long as the fir (Johnson 1992). Sprouters are unlikely to be killed by fire unless growing on shallow soil or organic soil, which may be heated to lethal temperature or consumed during the fire.

Wind is more likely to topple larger trees than smaller ones. Frelich and Lorimer (1991b) found that large trees (≥46.0 cm dbh) are 1.5 times more likely to topple than are mature trees (26.0–45.9 cm dbh), which in turn are 1.5 times more likely to topple than are pole-sized trees (11.0–25.9 cm dbh). There are three reasons for this. First, the larger a tree's diameter is, the less flexible the trunk will be. If a tree trunk is able to bend in the wind, the force is uniformly distributed over the length of the trunk. With a stiff trunk, all the force of the wind on the entire tree is applied at the base of the tree, which can result in uprooting or snapping off at a weak point (Petty and Worrel 1981, King 1986). Second, large trees are much more likely to have decay of some sort and this decay is most likely to be located near the base or in the root crown, causing a weak spot just where the force of the wind is applied. Some species of trees become completely hollow as they age, and the hollow tube of the trunk reduces the weight of the tree, so that less force is applied at the base during windstorms and the trunk continues to bend in the wind, despite very large diameters. American basswood is one such species that commonly attains a dbh of >1 m. Yet they don't blow down until the rate of outward expansion of the rot inside the trunk is faster than the growth of the trunk, which leads to thinning of the tube until the buckling point is reached. The third reason that large trees are more susceptible is that wind speeds are higher in the upper canopy where the crowns of large trees reside than in the understory (King 1986).

Soil type and rooting depth have a major effect on chance of wind-throw. Sandy soils provide the greatest stability, probably because of greater depth of root penetration, followed by loams, silts, clays, and peat (Fraser 1962). Root diseases such as fungal rots and variable depth to bedrock or water table blur this ranking of soil stabilities.

The force applied by the wind in trees increases linearly as wind speed increases from 'normal' velocities of 10–20 km/h to 100 km/h (Banks 1973, Frelich personal observation). The tree simply becomes more streamlined as the wind increases. First, leaves turn so that their sides are in line with the wind, minimizing their surface area. Next, small branches bend in such a way that twigs begin to line up in front of each other. Finally, larger branches bend and become streamlined in the wind so that the entire outline of the tree's crown presents a minimal surface area to the wind. At some point, however, no more streamlining can be accomplished, and increasing wind speed exerts a force that increases in proportion to the square of the wind speed after that. Sugar maple trees have the amazing ability to fold up their crown in high winds, so that a tree 30–40 cm dbh, with a crown diameter of 5–6 m on a calm day, pre-sents a profile approximately 2 m wide during winds of 110–120 km/h (Frelich personal observation). Some conifers can reduce their surface area by 45% in a wind of only 40 km/h (Banks 1973).

Many trees rock back and forth during long-lasting storms, such as cyclonic winds. Some winds may have a resonance effect where the sway of an individual tree is maximized when periodicity of gusts and the tree's sway correspond. This sway can also be damped by the soil/root complex, contact with other tree crowns, and aerodynamic effects of the tree's crown (Blackburn et al. 1988). Often, these swaying trees fall after many small roots break or a weak spot in the trunk is slowly exaggerated over several hours until the tree falls. Most trees fall in the same direction as wind movement, but sometimes there are surprises as when a particu-larly large gust is in phase with forward sway of the tree and suddenly lets up when the tree sways backwards. I have almost been hit by such back-wards falling trees while doing field work. During windy days, one always looks into the direction of the wind to spot falling trees and jump out of the way – a strategy that does not always work.

Blackburn and Petty (1988) calculated critical wind speeds for tree failure of 270 km/h, 170 km/h, and 184 km/h for sitka spruce planta-tions with 3 m spacing that were 7 m, 11 m, and 18 m tall, respectively. These values correspond closely to those that field researchers have observed to cause high levels of tree mortality in natural forests. Experience of several researchers is that in eastern North America winds of 100–120 km/h cause little tree damage, only taking out very large trees with basal decay. Winds 140–180 km/h cause partial destruction of the forest canopy, and winds >200 km/h cause canopy destruction by removing two-thirds or more of the stand basal area in mature or old-growth stands (Stoeckeler and Arbogast 1955, Dunn *et al.* 1983, Canham and Loucks 1984, Foster 1988b, Boose *et al.* 1994).

Optimum fire rotation periods for tree species
There is a balance between intervals between stand-killing fires and species composition. To perpetuate fire-adapted species, fires must not occur before the species is old enough to reproduce, but must occur before most individuals die of old age, leaving the stand without seed source, or before the species is replaced by late-successional species. Sprouters are the only species that can tolerate very short intervals between fires, since they can survive underground. Most conifers cannot sprout, but jack pine and black spruce have canopy-stored serotinous seed banks that are produced at relatively young ages (Figure 2.7). The long-lived conifers without canopy-stored seedbanks require moderately long rotation periods of a century or two for optimum development. There is no forest type in eastern North America that is documented to have a

Figure 2.7. Closed cones that open after fire (serotinous or bradysporous cones) predominate in the Lake States jack pine population. Photo: University of Minnesota Agricultural Experiment Station, Dave Hansen.

Table 2.1. *Natural rotation periods for stand-killing fire in cold-temperate forests of eastern North America*

Rotation period (yr)	Location	Forest type	Reference
50	BWCAW, Minnesota	Jack pine–black spruce	Heinselman 1973, 1981a,b
80	BWCAW, Minnesota	Aspen–birch–fir	Heinselman 1973, 1981a,b
80–170	Lower Michigan	Jack pine	Whitney 1986
150	Itasca, Minnesota	Red and white pine	Frissell 1973
130–260	Lower Michigan	Red and white pine	Whitney 1986
170–340	Lower Michigan	Mixed pine–oak	Whitney 1986
180	BWCAW, Minnesota	Red and white pine	Heinselman 1973, 1981a,b
175–300	Algonquin, Ontario, Canada	White pine–aspen	Cwynar 1978
800	NE Maine	Spruce–fir–beech–maple	Lorimer 1977
1400–2800	Lower Michigan	Hemlock–maple	Whitney 1986
2800–4500	Upper Michigan	Hemlock–maple	Frelich and Lorimer 1991a

natural rotation period between 300 and 800 years (Table 2.1). This is probably due to fuel-type–fire feedbacks. When the rotation period is long enough to allow species of low flammability such as hemlock or sugar maple to dominate the system, the species themselves lengthen the rotation period by creating fuels with low flammability. Thus, tree species in the Lake States fall into two groups: (1) fire-dependent species that become abundant when fire rotations are 300 years or less; and (2) non-fire dependent (usually fire sensitive) species that become dominant when rotations periods are more than 500 years (Figure 2.8). One species, black spruce, can fit with either group. It has the unique ability among Lake States species to reproduce well with short or long fire rotations.

Disturbance and succession

Four general principles on disturbance and succession emerge from the literature on temperate forests. Again, these are given as background in

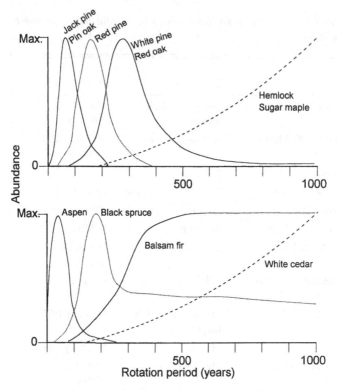

Figure 2.8. Relationship between rotation periods for stand-replacing fire and abundance of some important tree species in the Lake States. The *y*-axis is a unitless scale of abundance and each species is assumed to reach a maximum abundance at some rotation period. After Frelich (1992), and synthesized by author from field experience and many other sources.

this introductory chapter on properties of disturbance regimes, and will be expanded upon in later chapters:

- Crown fires set succession back when they burn a forest dominated by mid-successional or late-successional species. They do not change the forest successional state if the forest is already dominated by early-successional species (Dix and Swan 1971, Heinselman 1973, Frelich and Reich 1995a).
- Surface fires may set succession back when they burn forests dominated by late-successional species, but leave successional status the same when they burn a forest dominated by early-to-mid-successional species (Figure 2.9, Bergeron and Brisson 1990, Frelich and Lorimer 1991a, Frelich 1992, Clark and Royall 1995).

Figure 2.9. Surface fire in red pine forest: an example of surface fire preventing succession to balsam fir. Left, old red pines with balsam fir underneath. Right, red pines several years after a surface fire removed invading fir without killing the red pine. Photos: University of Minnesota Agricultural Experiment Station, Dave Hansen.

- Windthrow and harvesting leave the successional state the same in a forest dominated by late-successional species, but advance succession when they occur in forests dominated by early-to-mid-successional species (Abrams and Scott 1989, Abrams and Nowacki 1992, Frelich 1992, Tester *et al.* 1997, Frelich and Reich 1999).
- Herbivory by deer and moose retard the rate of succession when browsing in forests dominated by early-to-mid-successional species. They convert the forest to a lower tree density when browsing in forests dominated by late-successional species. Exceptions to these generalizations

occur depending the browser's species preferences (Watson 1983, Frelich and Lorimer 1985, Brandner *et al.* 1990, McInnes *et al.* 1992).

Disturbance–disturbance interactions

Wind

Wind creates burnable fuel of all sizes and time lags. Total canopy wind-throw over thousands of hectares can create conditions that will support large, severe fires in forest types that may not be flammable in a live con-dition (Lorimer 1977, Frelich and Lorimer 1991a).

Sequential windstorms interact with one another. One windstorm removes susceptible trees so that there may not be any trees available to blow down during the next windstorm. The period of lack of susceptible trees depends on the severity of the last windstorm. A 90 km/h windstorm may only remove a few trees with advanced trunk rot and within a few years there will be more trees with advanced rot that are susceptible to blow down by the same wind speed. A total canopy blowdown will leave a stand non-susceptible to blowdown of any severity for several decades (Frelich and Lorimer 1991a). Thus, the interaction among sequential windstorms and the vegetation partially regulates the rate of blowdown (Runkle 1982).

Finally, windstorms create downed trees, the twigs of which may be edible to browsers, and gaps in which small seedlings and saplings reach-able by deer can grow.

Fire

Fire consumes fine fuel but generally creates large-diameter dead fuel because most of the stems are killed but not consumed, even in severe crown fires. One only need wait for new fine fuel to grow in the forest to have the mixture of fine and large-diameter fuels necessary to support the next fire – a process that is likely to take only three to five years.

Fire makes injured live trees more susceptible to blowdown and insect infestation, and both removes some insect species and makes new habitat for others (McCullough *et al.* 1998). Severe fire creates a lot of dead trees which are not likely to blow down within the first decade after fire because there is little surface area for wind to push on. These standing snags are habitat for insects. The snags also eventually fall on the ground where they become coarse woody debris that can influence the species composition of the future stand.

For systems with a short rotation period for severe fire, most stands will not get old enough for wind to become a major form of disturbance. For example, a 50-year rotation for stand-killing fire in boreal jack pine would leave nearly two-thirds of the landscape covered with stands less than 50 years of age. Self-thinning is the major cause of tree death in young even-aged stands, rather than windfall-caused gaps.

Herbivores

Deer and moose can create a 'fuel gap' in the vertical structure of the forest. Deer and moose can clear out a zone up to 2 m or 3 m high, respectively. In this height stratum there may be little fine fuel or ladder fuel. The barren forest floor cannot sustain surface fires, and a combination of very low intensity surface fire and lack of ladder fuel can reduce the chance of crown fire.

Deer may change the successional course of development after severe windstorms. Normally in forests such as hemlock–hardwood there is a dense layer of suppressed seedlings on the forest floor that will be released after windthrow, and other tree species have little chance of competing with this dense seedling layer. However, if that seedling layer is removed by deer, then the forest may be left open to invasion by other species after wind knocks down the overstory.

There is also a deer–moose interaction. More deer in an area means fewer moose, because deer carry a brainworm (*Parelaphostronglyus tenuis*) that is lethal to moose. An increase in deer leading to a decrease in moose would have successional implications because of differential preferences between the two herbivores; this situation would be good for future prospects of balsam fir and paper birch, but bad for white pine, hemlock and cedar.

Deer and moose eat their preferred seedlings after fires and logging, altering the course of succession that the fire or logging alone would normally cause. These browsers may not eat all seedlings after severe blowdown because the slash limits deer movement and physically protects individual seedlings for several years (Peterson and Pickett 1995), which may be enough time for them to reach the no longer browsable height of 2–3 m.

Lastly, insect defoliators create burnable fuel of all sizes by killing trees and weakening trees that are not killed, making them more susceptible to blowdown in the long run. The short-term effect can be the opposite; within a few weeks of defoliation the trees have fewer leaves, and therefore are less susceptible to blowdown.

A classification of forest disturbance regimes

With several major disturbance types, each of which could occur with many different levels of severity, different frequencies, different sizes, and all the types interacting on a forested landscape, the problem of classification into important disturbance regimes may seem hopelessly complex. Examination of the literature, however, shows that suites of disturbance types and characteristics hold together in some cases. Also, to a great extent, fire overrides the other disturbance types, so that windstorms and herbivory by mammals and insects respond to the fire regime. Therefore, here is a cross-classification of fire frequency for light surface fires and stand-killing fires (whether crown fires or severe ground fires). An interesting pattern results (Table 2.2):

Regime I. Frequent light surface fire (Table 2.2, upper right). There are many forest (as well as woodland and savanna) types in this category (Heinselman 1981a, Kilgore 1981). These types all have open, park-like understories dominated by graminoids and canopies that are not dense enough to carry crown fires (at least not unless fire suppression leads to thick understory tree growth).

Regime II. Frequent fires of all types (Table 2.2, upper left). This regime leads to scrubby forest such as pine barrens (Curtis 1959). To sustain frequent low and high severity fires, the vegetation must respond with rapid re-establishment of dense brushy growth to facilitate the next fire.

Regime III. Complex fire regimes (Table 2.2, middle). These forests are dominated by mid-successional species that cannot reproduce if severe fire is too frequent, because they don't have canopy-stored serotinous seeds, bear seeds at young ages, and/or possess post-fire sprouting ability (Frissell 1973, Heinselman 1981a, Kilgore 1981). The dominant tree species also cannot compete with late-successional, shade-tolerant invaders. Thus, the combination of moderately frequent surface fires that rid the understory of shade-tolerant invaders, and some severe fires, but not too often, is their only possible niche.

Regime IV. Frequent severe fires (Table 2.2, lower left). These forests are characterized by widespread occurrence of young, dense, even-aged stands that seldom reach late-successional stages (Dix and Swan 1971, Johnson 1992). Many northern conifer forests in regions with occasional drought, such as the central North American boreal forest, fall in this category. This is also the domain of canopy-stored serotinous-seeded species, and some sprouters, such as aspen.

Table 2.2. *Cross-classification of disturbance by frequency of light fire and stand-replacing fire*

Light surface fire	Crown fire/severe surface fire			
	Frequent (25–100 yr)	Infrequent (100–500 yr)	Rare (500–1000 yr)	Very rare (>1000 yr)
Very frequent (<25 yr)	Jack pine sand barrens and aspen parkland (Midwest USA and central Canada)		Ponderosa pine (Rocky Mountains) Southern pines (southeast USA)	Bur oak savanna (Midwest USA) Giant sequoia (California)
Frequent (25–100 yr)		Red–white oak and white/red pine (Midwest USA)		
Infrequent (>100 yr)	Boreal forest (Central Alaska, NW Territories, Canada) Jack pine–black spruce and spruce–fir–birch (BWCAW, Minnesota)	Boreal forest (Quebec, Canada and northern Scandinavia) Black spruce peatlands (Minnesota) Lodgepole pine (Yellowstone) Douglas-fir (Rocky Mountains)	Douglas-fir, western hemlock, sitka spruce (USA Pacific Coast)	Sugar maple–basswood (Minnesota) Sugar maple–hemlock (Michigan, Wisconsin and northeastern USA)

Regime V. Moderately frequent severe fire (Table 2.2, lower middle). Forest types composed of long-lived species adapted to fire, but occurring in regions where severe droughts are not quite as common as the previous group (e.g. Cogbill 1985, Foster and King 1986, Bergeron *et al.* 1998).

Regime VI. All fire types infrequent (Table 2.2, lower right). Forest types in this regime have different dynamics from all the other forest types shown on the table. Wind becomes the dominant large-scale natural disturbance force and wind, insects and disease combine to cause small-scale gap-phase dynamics. The species strategies for perpetuation are also dramatically different from all those in the other regimes (e.g. Whitney 1986, Frelich and Lorimer 1991a). This is the domain of shade-tolerant species that have little in the way of seedbanks (canopy stored or soil stored), but that survive disturbance in the form of a seedling bank, or as resprouts from the stump.

Disturbance regimes on the landscape

The interaction between frequency of drought and physiography determines the spatial relationship among landscape units with the six disturbance regimes described above. A climatic gradient from dry to wet on uniform physiography would typically lead to a gradient in forest disturbance regime types that follows the order I–VI above. Frequent light surface fires (Regime I) occur at the edge of grasslands that burn more than once a decade, and fuels never reach high loadings so fire intensity remains low. With a slightly more moist climate, fires may sometimes occur almost annually and then skip a decade or more, depending on the sequence of wet and dry years; hence Regime II with light surface fires alternating with more severe fires. Next come the complex fire regimes (Regime III), in systems where surface fires are not quite frequent enough to prevent successful recruitment of young tree cohorts, and severe crown fires are not very frequent either. It is in this regime that the trade-off between surface fires and crown fires is evident; it is in some sense a compromise between the frequent light fire regimes and the infrequent severe fire regimes. In Regime IV, fires are infrequent enough that massive fuel loads build up, resulting in very severe fires when an infrequent drought occurs. Droughts and resulting severe fires are rare in Regime V, and succession to late-successional species between fires becomes more likely than in Regime IV. Finally, in Regime VI the climate and soil combination is so moist that fires are extremely rare, and the species that grow there have fuel characteristics that limit fire intensity unless extreme fuel conditions occur, such as windfall slash.

Because climatic gradients that are large enough to allow the full sequence of Regimes I–VI to exist also occupy large spatial domains, differences in physiography are likely to interfere with the development of an ideal disturbance regime gradient. Bodies of deep loamy soil allow existence of disturbance regimes one or two numbers higher than the surrounding regimes; and conversely areas of shallow or coarse sandy soil allow existence of regimes one or two numbers lower than the climate alone might indicate. On landscapes with very diverse topography and landforms, these features may almost override a regional climatic gradient. Northeastern Minnesota provides an example of a climatic gradient that overlies diverse landforms that greatly modify the expected gradient of disturbance regimes (Figure 2.10). The Saganaga Batholith, an area of extremely shallow soil over granite in the Boundary Waters Canoe Area Wilderness, frequently has dry fuels, and lightning ignitions are common, even though lightning frequency is no higher than elsewhere in the landscape. Hence, this area is classified as Regime IV, even though it is in a moist climate in the extreme northeastern part of the map (Figure 2.10). Disturbance Regime VI is able to express itself in a few areas with deep loamy soil fairly near the prairie–forest border (Figure 2.10). Near Lake Superior, one would expect only Regime VI to exist, but areas in Regimes III and IV occupy most of the area (Figure 2.10).

Summary

This chapter gives background information on disturbance necessary to understand later chapters. The definition of the disturbance regime (a description of the type, size, intensity and frequency of disturbance) starts the chapter. Next two very important basic concepts were explored: rotation period (time required to disturb an area equivalent to the whole landscape of interest) and difference between intensity (a measure of physical energy of a disturbance) and severity (the impact on plants via degree of mortality). Then the mechanics of the three most important natural disturbance types in temperate forests were considered: wind, fire, and herbivory. These three disturbance types have many interactions among themselves and with the forest. Near the end, the chapter goes full circle and returns to the disturbance regime definition. Armed with the information on frequency, size, type, and intensity/severity of disturbance from the middle of the chapter, it is possible to outline six major

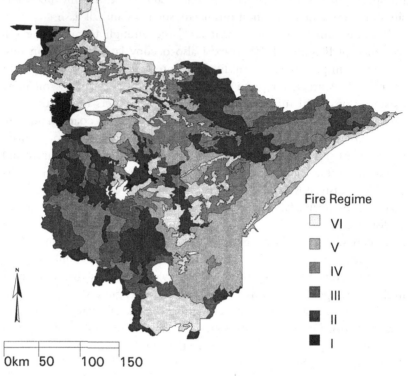

Figure 2.10. Distribution of fire regimes in northeastern Minnesota. Map modified from Dave Shadis, U.S. Department of Agriculture, Forest Service, Chippewa National Forest, Minnesota. Note that if climate were the only influence, the regimes would fall in order from I to VI from southwest to northeast across this map.

disturbance regimes that fall within a scheme of fire frequency cross-classified according to frequency of stand-killing fires and frequency of light surface fires. The life-history characteristics of the dominant tree species with regard to shade-tolerance and mechanism for surviving or regenerating after fire, and the other disturbance types, work together with the cross-classified fire regimes to form the complex disturbance regimes. Thus, the simple definition of a disturbance regime given at the beginning of the chapter can be reconciled with the complex multivariate nature of disturbance as it occurs on the ground. The disturbance regimes can also be mapped on the ground because they fall along a climatic gradient from dry (disturbance regime I) to wet (disturbance regime VI). Finally, the chapter showed that there is an interaction between the dry-to-wet gradient of disturbance regimes and soil type. For example, sandy or shallow soils can cause a landscape patch to show disturbance Regime I or II dynamics when it would be Regime III on a loamy soil or Regime V on a silt loam, with identical climate in each case.

3 · Sampling and interpretation of stand disturbance history

Introduction

The previous chapter provided general background information on how disturbances work in the forest. This chapter shows how to detect and measure the impact that disturbances discussed in Chapter 2 have upon the forest at the individual tree and stand scale (1–10 ha). Chapter 4 follows with a synthesis of stand dynamics in the Great Lakes forests, obtained using the methods presented in this chapter. Thus, this pair of chapters (3 and 4) presents methods for studying stand dynamics and then the results of application of those methods to the Great Lakes forest. Chapters 5 and 6 form a similar pair of methods/applications but at the landscape scale.

Here I start with discussion of the different types of evidence on stand disturbance history, then proceed to show how to use such evidence to construct a *disturbance chronology* which chronicles the occurrence of disturbance and its impact on stand structure for the last few centuries. Much of the chapter is devoted to interpreting tree rings for stand history and dealing with the various problems that are inherent in these methods. The importance of choosing a sampling scheme (i.e. the all-important question of which trees to core) that matches the objectives of a given study is another important topic discussed below.

Use of stand data for interpretation of stand history

Several lines of field evidence are available to those investigating stand history. They provide different levels of detail, time resolution, and types of information about stand history (Table 3.1). Much of the chapter will concentrate on tree-ring evidence, because that type has the widest availability and greatest utility for fine resolution of temporal and spatial disturbance processes in temperate zone forests.

Table 3.1. *Properties of different lines of field evidence when used to interpret stand history*

Type of evidence	Availability	Type of information	Attribute			
			Temporal resolution (yr)	Spatial resolution (m)	Length of record (yr)	
Fossil	Spotty; requires presence of lakes, bogs, most available in glaciated regions	Presence or absence of tree species; sometimes relative abundance; occurrence of fire	Medium–poor: 10s–1000s	Poor: 100s–1000s	Very long: centuries to many millennia	
Historical	Spotty; requires intact written record	Variable, but may include timing and severity of storms, fires and cutting	Medium: 10s–100s	Poor–medium: 100s–1000s	A few years to millennia	
Physical	Nearly universal	Tip-up mounds	Medium: 100s	Medium–high: 1–100	A few centuries	
Air-photo	Common in recent years	Species and developmental stage	Medium–high: 1–10s	Medium: 10–100	Several decades	
Compositional	Universal	Presence of certain species indicate disturbance	Medium: 10s–100s	Medium–high: 1–100	A few centuries	
Structural	Universal	Size of trees is related to time since disturbance	Medium: 10s–100s	Medium–high: 1–100	A few centuries	
Canopy gap	Universal	Composition of the 'next generation' of trees	Medium–high: 10–100	High: 1–10	A few decades	
Tree ring	Universal in temperate zones	Total age, scars and/or release indicate time since disturbance	High: 1–10s	Very high: 1–10	Mostly 1–2 centuries; rarely a few millennia	

Fossil evidence

Fossil pollen and plant parts such as seeds and conifer needles preserved in sediments that can be radiocarbon dated have allowed reconstruction of tree migration over thousands of years and large regions (Davis 1981). Reconstruction of major changes in forest-community composition in response to climate change and disturbance at a given location can also be accomplished with such data. There are conditions under which these analyses can be used to investigate stand history. When small, sediment-filled hollows are present that are covered with forest canopy, as much as half of all pollen falling on the surface of the sediment may have traveled through the trunk space, rather than above the canopy. This means that a strong local signal may exist at the scale of 1–2 ha of forest (Calcote 1995, 1998).

There are several well-known problems with the fossil pollen data. For example, differential preservation sometimes makes the analysis of forests with *Populus* difficult. One cannot always differentiate key species in the same genus. Paper birch and yellow birch are indicative of different types of disturbance (fire versus wind, respectively), but their pollen is indistinguishable. There are great problems with calibrating the amount of pollen in a given stratigraphic layer to the abundance of trees in the vicinity. The relative pollen production of each species and different distances that pollen of each species travels makes this an almost intractable problem. Despite these difficulties, pollen analysis is a valuable tool, when looking at differences in composition that take place over hundreds or thousands of years. Switches in forest type at the stand scale, when they occurred and whether they were gradual or abrupt, have been analyzed in hemlock and sugar maple forests of the Lake States (see Chapter 6, 'Sylvania case study').

Charcoal can indicate the occurrence of fires in the vicinity. There are problems with source area, however, which makes it hard to pinpoint a return interval for fires. The use of charcoal in studies of forest history has generally been limited to changes that occur over thousands of years (e.g. Swain 1978, Clark and Royall 1995). However, when one is looking at forest development over a more limited geographical area (a local stand) or a much shorter time scale (the life of the current generation of trees), techniques for dating charcoal, such as varved lake sediments and carbon-14, are of more limited use. Little is known about how long charcoal persists in the forest floor, but buried charcoal is very persistent (Buckley and Willis 1970). The presence of charcoal in a forest floor therefore does not necessarily indicate that the current stand originated after fire. Henry and

Swan (1974) found charcoal which they concluded from other evidence was 300 years old, yet the stand was only about 35 years old, having originated after the 1938 New England hurricane. Carbon-14 analysis could be used to distinguish between ancient and modern fires, but the time resolution is not fine enough to be very useful if the objective is reconstruction of forest history over the last century. Even with varved lake sediments, charcoal deposition events can be dated exactly, but they cannot always be matched up in a one-to-one correspondence to fires. This is because heavy rain storms may wash charcoal into the lake on multiple dates after one fire.

Historical records

Often there are records of severe storms or forest fires; sometimes the exact stands that were hit can be identified. For example, Henry and Swan (1974) knew that their study area had been hit by the 1938 New England hurricane. They were also able to cite Channing (1939) for dates of earlier hurricanes that may have caused one or more of the disturbances on their study area. Managers of parks and forests frequently gather historical information about their management area, as well as keep a record of storm damage, fires or timber cutting that takes place. Thus, Lorimer and Krug (1983) were able to verify the even-aged condition of many of their stands from experimental forest records. Frequently surveyors or early settlers, foresters and naturalists give useful information about specific areas. Examples are Roth's (1898) survey of the forest conditions and Irving's (1880) account of a large windfall 40 miles in length, both in northern Wisconsin. Finally, weather records are useful sources of historical information. Accounts of major storms and droughts extend back to the early 1900s in the journal *Monthly Weather Review*.

Physical

This evidence in temperate forests is mainly limited to windthrow mounds, sometimes called tip-up mounds, or pit and mound topography. If a mound is associated with a rotten log a date of formation can often be established. In this case it may be possible to establish the date by careful excavation around the mound and determination of ages of living and dead trees near it (Stephens 1956, Henry and Swan 1974). By contrast, analysis of windthrow mounds is of limited value for determination of disturbance severity; the number of mounds on a per-hectare basis does not correspond to severity of the last disturbance in many cases. Personal observation of cored trees growing on windthrow mounds in Upper Michigan indicates that many windthrow mounds remain visible for 200 years or more. Mounds are a preferred site for germination and survival of new trees, whether they are formed by single treefall gaps or a stand-leveling disturbance. Since trees are uprooted throughout the life of a stand and mounds may persist from a previous stand, it is difficult to establish a link between the number of mounds present in a stand and the date and intensity of a wind-caused disturbance. Thus, the study of these mounds is more useful for establishing microsite preferences for understory plant species and tree seedlings (Beatty 1984, Beatty and Stone 1986), than for analyzing stand disturbance history.

Air photo interpretation

Sequential air photos are especially valuable for verifying and interpreting successional and stand development sequences. Often, stereo pairs of air photos are available, and many keys allow one to identify tree species. If one can obtain two or three sets of photos of the same area taken over several decades, then two sets can be viewed at the same time on a zoom transfer scope, which allows for adjustment of any scale differences that may occur between two sets of photos. Using this technique, it is possible to verify whether chronosequences are a valid way to study successional pathways. For example, Frelich and Reich (1995b) found that jack pine stands that were 30 years old on air photos taken in 1934 looked the same – that is they had the same tree species composition and canopy texture – as stands that in the 1990s are 30 years old. This verified that, at least for the last several decades, there was a stable chronosequence, and the chronosequence technique would be valid for field studies of succession.

Compositional

The species composition of a stand may contain some information about disturbance history. There are few direct relationships between disturbances and the succession of tree species which occurs after a disturbance. In recent years, many investigators have abandoned the view that the process of succession is so homogeneous that the approximate age structure of the forest can be deduced from it (Cattelino et al. 1979, Oliver 1981, Franklin and Hemstrom 1981). It is tricky to relate age and species composition except in cases where a thorough ecosystem classification has been done, the growth characteristics of the tree species have been characterized for the ecosystem type, and one is sure that events such as climate change have not influenced these factors during the lives of the current generation of trees. In the hemlock–hardwood forest only paper birch–aspen stands can reliably be associated with disturbance. Such stands occur with high frequency after intense fires (Lorimer 1981). Even-aged stands of paper birch–aspen mixed with other species are probably of fire origin, especially if verified by datable fire scars on survivor trees. It should be kept in mind, however, that the fire may have burned in windfall slash. Thus, the frequency of windfall may be underestimated if all paper birch–aspen stands are considered of fire origin only.

Structural

There is a relationship between stand structure and disturbance history, but it has limitations. Diameter–frequency distributions (referred to hereafter as diameter distributions) are sometimes useful when used in conjunction with age data. To avoid complications caused by lumping different even-aged stands or species with different growth rates, diameter distributions should be plotted separately for the leading dominant species on a single homogeneous stand (Hough 1932, Meyer 1952, Oliver 1978). Also, the shapes of diameter distributions for canopy trees only (those receiving direct sunlight on the top of the crown) are much easier to interpret than when all crown classes of trees are considered together. With these considerations taken into account, all-aged forests commonly have a steeply descending monotonic diameter distribution (De Liocourt 1898, Hough 1932, Tubbs 1977). By contrast, an even-aged stand usually exhibits a unimodal diameter distribution (Meyer 1930, Hough 1932, Lorimer and Krug 1983). Unbalanced multi-aged stands and even-aged stands, however, cannot reliably be separated on the basis of diameter distribution without tree age information.

Another useful form of structural evidence is a diameter-exposed crown-area distribution (ECA distribution). To obtain the crown-area distribution of a stand, the cross-sectional area of the portion of the crown exposed to the sun of each tree is estimated (thus eliminating suppressed trees and areas of crown overlap). These areas are then summed for all trees within each size class. The resulting distribution represents the proportion of the total exposed crown area (of the stand) occupied by each size class. In a perfectly balanced all-aged forest new patches would, by definition, be created at a constant rate, resulting in a mosaic of groups of trees of all different ages (Lorimer and Frelich 1989). Trees in each age class would have unimodal diameter (and crown-area) distributions. The total crown-area distribution would then be the sum of widely dispersed unimodal distributions of the age classes. The principal advantage of crown-area distributions over diameter distributions is that different densities which occur in various size-classes are equalized. For example, a gap created by the fall of a large tree may contain 50 saplings, so that 50 saplings occupy the same area as one large tree. In cases like this, density exaggerates the importance of small trees in a stand.

Tree replacement in 'gaps'

Gaps can be sampled across a forest stand by walking transects and measuring length, width, date of formation, and composition of the gap-making trees and replacement trees in all gaps that were intersected (Runkle 1981, 1982). If a large sample size is obtained, a simple Markov succession model can be constructed, using the proportions of basal area among species of gap-making trees (the 'current forest generation') and the proportion of basal area or density among replacement trees (the 'future forest generation'). In addition, the proportion of area in the stand going through gaps per unit time can indicate the canopy turnover rate. For example, 10% turnover per decade would seem to indicate a 100-year canopy turnover.

Tree age data

The last line of evidence for analysis of stand history is the use of age and growth-rate data obtained by increment coring. Distributions of total tree ages are useful but do not always give a direct indication of stand disturbance history. Age distributions are unambiguous only when a stand is strictly even-aged (Lorimer 1985). Such a stand could be created by an intense crown fire that kills understory vegetation, or by catastrophic windthrow in a stand with no understory (i.e. the stem exclusion stage, Oliver 1981). All of the trees in the stand would originate as new seedlings. However, the same type of disturbance could also create a unimodal or descending monotonic age distribution if recolonization following a disturbance is slow (Franklin and Hemstrom 1981). Broadly unimodal age distributions can also be created by a series of disturbances (Oliver and Stephens 1977), a heavy disturbance which allows understory trees of various ages to survive (Henry and Swan 1974, Lorimer

1983a), or interruption of seedling recruitment (such as by deer brows-ing, Anderson and Loucks 1979). Finally, an apparently 'all-aged' distri-bution may develop after a stand-leveling disturbance. Consider the case of a hardwood stand which had the canopy removed by a windstorm 100 years ago. There could be small suppressed trees up to 100 years old which survive. In addition, new seedlings would be established on the forest floor in the post-disturbance stand. The range of ages in such a stand could be 200 years.

The most important evidence of stand history, also obtainable from increment cores, is the pattern of growth increment. When large trees are windthrown, other trees which were formerly suppressed or crowded by the windthrown trees usually show an abrupt and sustained increase in radial growth. So long as possible errors (such as missing rings and inter-ference with normal growth caused by climatic anomalies) are taken into account, such releases from suppression can clear up ambiguities present in age distributions (Lorimer 1985). Understory trees of various ages which are released from suppression by a massive windstorm will all show release about the same date, making releases from suppression a more accurate indicator of disturbance intensity than age distributions.

However, care must be taken in interpretation of disturbance intensity from increment cores. There is not necessarily a one-to-one correspon-dence between the proportion of trees showing release and the propor-tion of the canopy destroyed by a disturbance (see 'Construction of stand disturbance chronologies' below). A tree that has been released may later be overtopped by gap closure or by other faster growing trees (Oliver 1978, Bicknell 1982, Hibbs 1982). If such trees remain overtopped, the original disturbance may not be detected if understory trees are not included in the analysis. By contrast, trees some distance from a newly formed gap may respond to increased light (Canham 1985), resulting in an overestimate of disturbance intensity. Growth patterns are also useful for determining the proportion of trees which enter a stand as new seed-lings after a disturbance with no period of suppression. These trees grow relatively rapidly from the start and show a characteristic pattern of stead-ily declining radial growth with age. It should be possible to distinguish those sample trees that were in the canopy from the start, based on growth patterns, and to combine this information with releases from suppression to obtain a disturbance chronology for a given stand of trees. Much more detail is given on this in the following section.

Fire scars have been used extensively to date fires in conifer forests. Arno and Sneck (1977), Stokes (1981) and Madany *et al.* (1982) discuss techniques for selecting sample trees and problems involved in determining dates from fire scars. Errors of plus or minus several years caused by missing, false or indistinct rings may occur in a particular tree (Stokes 1981). Since these errors may result in assigning more than one date to a single fire, the dates should be corrected using the principles of cross dating (Fritts 1976). In eastern North America, analysis of fire scars is often difficult due to the decay of wood around the scar which occurs in the humid climate. Although fire scars are occasionally found on old specimens of hardwoods such as sugar maple and oaks, relatively little analysis of fire history has been done in hardwood stands as compared with conifer stands. Some exceptions are studies of fires scars on oaks (Buell *et al.* 1954, Abrams 1992, Abrams and Nowacki 1992). In contrast to hardwoods, some eastern conifers retain fire scars quite well. Successful analysis of multiple fire scars has been done on red pine, eastern white pine and white cedar, (Spurr 1954, Heinselman 1973). Hemlock trees with fire scars dating from a fire episode in the 1920s–1930s were used to outline the extent of the fire in the Porcupine Mountains (Frelich and Lorimer 1991a).

Construction of stand disturbance chronologies from tree-ring evidence

Three issues are at stake here: (1) how to use tree-ring measurements and analysis of radial increment patterns to determine when a tree entered the canopy (canopy accession date) and/or when it reached a certain size class; (2) sampling stand age structure, i.e. for exactly which trees do we need to know canopy accession date? and (3) how to combine the canopy accession dates from the sampled trees into a disturbance chronology for the stand.

Radial increment pattern analysis

Studies commonly reconstruct disturbance history of forests based on age of release from suppression or other growth patterns (e.g. Heinselman 1973, Henry and Swan 1974, Canham 1985, Glitzenstein *et al.* 1986, Foster 1988a,b, Lorimer and Frelich 1989, Deal *et al.* 1991, Dynesius and Jonsson 1991, Frelich and Lorimer 1991a, Frelich and Reich 1995b). Indeed, releases and growth trend over the life of a tree are very valuable for interpreting stand history, even though they are what dedrochronologists studying climate as recorded by tree-rings filter out of their data. In complementary fashion, most of the criteria discussed in this section are designed to filter out the climatic influence on tree rings, revealing the disturbance history. Often, the date at which a tree attained a certain size is important to reconstruct as well. For example, the size structure of a stand years ago may be important for postdicting the suitability of the stand as habitat for a wildlife species that utilizes trees of a certain size.

Interpretation of releases

A release is defined here as a large-magnitude, abrupt and sustained increase in radial growth (Figure 3.1). Releases occur in understory trees after removal of the canopy (release from suppression) or in canopy trees when one or more neighboring trees are removed (release from competition of neighbors). Shade-tolerant species are capable of exploiting very large changes in available sunlight, even when they are old and large trees, whereas intolerant trees are not. For example, Frelich and Lorimer (1991a) found many sugar maples and hemlocks with periods of extreme suppression ranging from a few decades to 300 years. Mid-tolerant species such as yellow birch and white pine generally can only sustain 10–50 years of suppression by shading, and intolerant species such as paper birch and red pine can only sustain short periods of about 10 years. However, intolerant species such as red pine can sustain very long periods of suppression (up to a century) due to overstocked stand conditions,

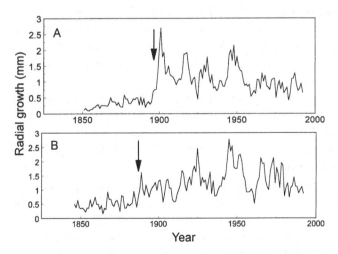

Figure 3.1. Examples of releases from suppression in radial increment patterns from red pines studied by Frelich and Reich (1995a). A, major release (≥100% growth increase); B, moderate release (50–99% growth increase).

where each tree is still receiving some direct sunlight on top of the crown (Frelich and Reich 1995b). Thus, releases in intolerant species do not necessarily indicate when they first entered the canopy, but rather when a stand-thinning event occurred.

Three terms from the definition of release given above – 'large-magnitude', 'abrupt', and 'sustained increase' – require further explanation. Criteria must be set up for these three terms which are reasonable in light of the species and region involved in a given study.

The magnitude of release for a given tree depends on how large a change in environment occurs and the size and species of tree. Percentage increases, rather than a fixed release threshold (i.e. 0.5 mm/yr for sugar maple in Canham 1985), have been used more widely. The reasons for this are that some understory trees would grow faster than any reasonable fixed threshold even before a release, whereas other canopy trees which have been released may not attain the threshold growth rate until many years later, if ever. In addition, older trees with large diameters may never attain a ring-width threshold that works for smaller trees.

In hemlock–hardwood forest, radial growth increases in saplings of two- to six-fold after release from suppression are well documented for sugar maple (Downs 1946, Eyre and Zillgitt 1953, Lorimer 1983a, Canham 1985) and for hemlock (Marshall 1927, Lyon 1936, Lorimer 1980). Dahir and Lorimer (1996) found that sugar maple saplings showed average growth increases of 92% and 144% when very small gaps (0–50 m^2) and medium-sized gaps (100–150 m^2) formed above them, respectively. Mid-tolerant species such as yellow birch do not respond to release quite as well (Eyre and Zillgitt 1953), but Godman and Marquis (1969) found an average 78% increase in radial growth of yellow birch saplings after a heavy crop-tree thinning, and many cases of 100% increase are evident in the cores collected by Frelich and Lorimer (1991a). Sapling and pole-sized canopy trees often respond to increased growing space, such as that from light crop-tree thinnings or group selection cuts with moderate release (Eyre and Zillgitt 1953, Stone 1975, Stroempl 1983). Such trees may be growing fairly rapidly before release and cannot be expected to increase radial growth as dramatically as a suppressed tree (Godman 1968). Canopy trees growing in extremely dense stands that are heavily thinned so that new growing space is opened up on more than one side of the crown may show 100% or more growth increases. This is true for young stands of tolerant or intolerant species and for mature tolerant species (Downs 1946, Godman and Marquis 1969, Frelich and Reich 1995b, Singer and Lorimer 1997).

How are we to use the complex data just presented for interpretation of stand disturbance history? Thresholds for increases in ring width need to be large enough to screen out responses to side light that do not directly represent a gap-forming event (i.e. minimize the 'false positives'), and they need to be small enough not to exclude trees that have responded directly to a gap (i.e. minimize 'false negatives'). Several researchers have weighed these issues and have used criteria of >100% and >50% increases in ring width after disturbance to indicate 'major release' and 'moderate release' for shade-tolerant species (e.g. Frelich and Lorimer 1991a, Frelich and Graumlich 1994). In these cases major release was defined as transition from understory (no direct sunlight) to a canopy position (crown receives direct sunlight), and moderate release was taken to indicate an increase in growing space for trees already in the canopy. An exception was made for cases where a moderate release is the only release for a given tree that is known to be in the canopy at the time of coring. In such cases, the release date is the most likely time at which the tree entered the canopy (e.g. Lorimer and Frelich 1989, Frelich and Graumlich 1994). Remember that these criteria create a reasonable balance between false positives and false negatives for the forests of interest. They may have to be adjusted to accommodate different forest types or different study objectives. For example, Nowacki and Abrams (1997) developed a method for determining the dates of canopy disturbance based on response of old canopy oak trees to decreased competition when one or more neighbor trees died. Their analyses were based on increases in growth that are higher than those that could reasonably be expected due to climatic effects during the years in question. A variety of growth thresholds was obtained for different trees and different decades over the last 300 years in Pennsylvania oak forests.

In addition to magnitude of growth increase, a certain degree of abruptness is also required to screen out patterns that may not represent a release. In a truly abrupt release there is a period of 1–5 years during which the >50% or >100% increase in ring width increase occurs. These criteria are consistent with data presented by Stroempl (1983), which show sugar maple responding to release within two years and reaching a peak radial growth rate 5–6 years after a thinning. An obvious point where the change from a relatively slow to a fast growth rate occurred is visible. One can always adopt two or more categories of release such as 'abrupt' and 'gradual' to provide some sensitivity analysis of the criteria chosen to indicate release.

Finally, there is the question of how long an increase in growth should be sustained in order to be considered a release. Several studies have adopted criteria of 15 or 20 years of sustained slow growth before, and sustained high growth after a proposed release. There is a trade-off to be made between a relatively short period of sustained increase and a relatively long one. Use of a long period of sustained increase in radial growth would screen out patterns caused by short-term climatic fluctuations, responses to side light by understory trees next to a gap (Canham 1989), and response of understory trees to canopy defoliation by insects, which often results in 2–3 years of increased light. Use of a short period of sustained increase in radial growth would screen out trees which are overtopped after only a few years by faster growing trees in a gap, and minimize the effect of decreasing radial growth that occurs as trees increase in diameter. Observations of radial growth on the Upper Michigan study areas during the drought of the 1930s indicate that trees on hilltops have a depression of growth rates for 4–10 years, while trees on lower slopes show almost no evidence of the drought. This was the most severe prolonged drought on record in the adjacent northern Wisconsin area (Lorimer and Gough 1988) and represents the most dramatic climatic event that must be separated from releases in the Great Lakes Region. Studies of hemlock radial growth in New Hampshire from 1860 to 1927 (Lyon 1935) and in New York from 1930 to 1970 (Cook and Jacoby 1977) reveal patterns of growth similar to those in Upper Michigan. No climate-related periods of slow growth longer than 10 years were reported in either study. Examination of Lyon's (1935) data reveals no events that would be interpreted as releases under the proposed 15-year criterion, even though there was a 10-year period of below-normal precipitation in New Hampshire during the 1860 to 1927 study period. Thus, using criteria of at least 15 years of slow growth before release and at least 15 years of rapid growth after release seems to be a reasonable way to screen out growth patterns that are anomalous from the disturbance ecologist's point of view. The 15-year pre-release criterion may have to be relaxed in cases where a tree shows a major release when it was less than 15 years old, so that there is not the possibility of 15 years of slow growth prior to the release event.

Analysis of early growth rates

Some canopy trees show no releases from suppression. In many cases such trees were already in a gap at the time of the earliest ring on their increment core, and it is desirable to have objective methods for classifying them as such. Providing objective methods for handling cores without

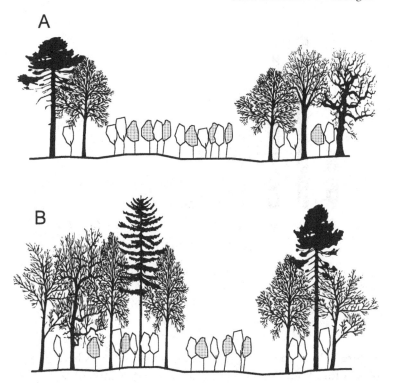

Figure 3.2. Two hypothetical forests illustrating the effects of forest structure on the chance of suppression given an observed sapling radial growth rate. Saplings with shaded crowns are growing faster than a threshold growth rate such that 50% of saplings in gaps exceed the threshold, but only 20% of suppressed saplings exceed the threshold growth rate. In case A, 1 of 7 (14%) of all saplings growing faster than the threshold are suppressed. In case B, 2 of 5 (40%) of all saplings growing faster than the threshold are suppressed. After Lorimer *et al.* (1988).

releases is the purpose of this section of the chapter (and the next section as well, see 'Other radial increment patterns' below). Shade-tolerant trees have a well-known ability to remain alive for many years in dense shade. Such trees persist in the understory with much slower growth rates than canopy trees overhead (Graham 1941a, Canham 1985, 1989). Therefore, it is reasonable to expect that one may classify the canopy status of a large tree at the time it was a sapling (i.e was it in a gap or suppressed?). A classification rule was devised to determine the probability that a sapling with a given growth rate came from the overstory distribution (Lorimer *et al.* 1988, Lorimer and Frelich 1989). A comparison of two hypothetical forests with different structural features illustrates how such a rule can

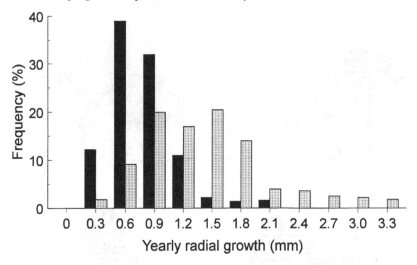

Figure 3.3. Distributions of radial growth rates for suppressed (black bars) and gap (shaded bars) saplings. All of the saplings are sugar maple 2–4 cm dbh and were growing in the hemlock–hardwood forest of the Porcupine Mountains, Michigan. After Frelich (1986).

work (Figure 3.2). In the example, 20% of suppressed trees have growth rates above a threshold value of 1.0 mm/yr, whereas 50% of saplings in gaps exceed this value in both forests. In a forest with a large area in recently formed gaps (Figure 3.2A), 1 of 7 (14.3%) of all saplings that are growing faster than 1.0 mm/yr are suppressed. In the other stand with less gap area, the probability that a fast-growing sapling is suppressed is much higher, 2 of 5, or 40% (Figure 3.2B).

These probabilities can also be used to determine the gap status of a tree – although it may be large and old today – at the time it was a sapling. To do this, the mean 5-year growth rate of each sample tree is measured at the point where it was a sapling (4 cm dbh for the examples here) and this growth rate is compared with contemporary growth rate distributions for suppressed saplings and gap saplings in the 4 cm dbh size class (Figure 3.3). If a sample tree has an observed growth rate, X, $\geq$ a specified growth rate, x, then the probability that the sample tree originated in a gap is equal to the ratio of the proportion of gap saplings exceeding growth rate x to the proportion of both suppressed and gap saplings exceeding this threshold (Lorimer and Frelich 1989):

$$P(\text{gap} \mid X \geq x) = \frac{G_x \times Q'_g}{(S_x \times Q'_s) + (G_x \times Q'_g)} \tag{3.1}$$

where:

$P(\text{gap} \mid X \geq x)$ is the probability that a tree was growing in a gap when 4 cm dbh, given a mean 5-year radial growth, X, at 4 cm dbh $\geq$ a specified threshold growth rate, x

G_x is the proportion of saplings in the 4-cm diameter class with growth rate $\geq x$

S_x is the proportion of suppressed saplings in the 4-cm diameter class with growth rate $\geq x$

Q'_g is the proportion of current canopy trees in the stand (or in stands with similar structure to that being analyzed) that were growing in gaps when 4 cm dbh

Q'_s is the proportion of current canopy trees in the stand (or in stands with similar structure to that being analyzed) that were suppressed when 4 cm dbh.

One can obtain G_x and S_x directly from the contemporary sapling growth rate distributions. Note that $Q'_g + Q'_s = 1.0$, and that they can be viewed as weighting factors for situations where the relative abundance of currently mature trees that were in gaps and those that were suppressed when they were 4 cm dbh are something other than 50:50. Therefore, we need only obtain one of them to get the other, and Q'_g can be estimated in a recursive fashion (Lorimer et al. 1988 present an example). The recursive procedure involves an initial estimate of Q'_g as the proportion of all trees that were sampled that have a growth pattern indicating gap origin (see 'Other radial increment patterns' below), plus all current canopy trees in the sample with growth rates at 4 cm dbh $\geq$ an initial arbitrarily chosen threshold, and then calculating an interim $P(\text{gap})$. The threshold growth rate is adjusted gradually until $P(\text{gap}) \geq 0.95$, establishing the minimum threshold for 95% confidence of gap status at 4 cm dbh (and at the height the cores were taken).

This early growth rate analysis should be developed separately for sites with different quality and stands with different structural types because growth rate distributions may vary. One is also assuming that growth-rate distributions are stationary over time (i.e. the growth rates were the same under the same conditions when mature trees were saplings as they are for present-day saplings).

If the assumptions are valid and the procedure has worked, the following formula should provide a good prediction of H_x, the proportion of all trees in the 'historic' data set of mature trees with growth rates $\geq x$:

$$H_x = (S_x \times Q'_s) + (G_x \times Q'_g) \tag{3.2}$$

When applied to the hemlock–hardwood forest of Upper Michigan, Lorimer and Frelich (1989) found that sugar maple and hemlock saplings 2–6 cm dbh had a 95% chance of growing in a gap if their annual radial increments averaged 1.5 and 1.3 mm/yr, respectively, in old growth stands. In younger stands dominated by pole and mature-sized trees, the thresholds were different: 1.2 mm/yr and 1.0 mm/yr for sugar maple and hemlock, respectively.

Other radial increment patterns
Other types of radial increment patterns besides releases are useful for interpretation of stand history. Many trees never go through a period of suppression, especially when seedling establishment occurs after severe crown fires and stand-leveling windthrow. These trees exhibit nearly flat, declining, or parabolic radial increment patterns (Figure 3.4). Nearly flat radial increment patterns (minimum 10-year average ring-width at least 70% of the maximum 10-year average ring width) and gently declining patterns lasting all or most of a tree's life have been found to be very common in even-aged stands and in large gaps within multi-aged forests, even for shade-tolerant species like sugar maple and hemlock (Hough and Forbes 1943, Frelich and Lorimer 1991a). If such trees are now in the canopy (at the time of coring), then they were probably in the canopy all their life. The logic of the situation is that it is unlikely that a tree would advance from suppressed status to free-to-grow status while showing a decrease (or at least no increase) in radial growth rate. The same reasoning was used by Oliver and Stephens (1977) in analyzing radial-increment patterns of a mixed species forest in New England.

A second radial-increment pattern frequently observed is parabolic (Figure 3.4). For example, Lorimer and Frelich (1989) found that most maples in a 40-year-old stand that originated after stand-leveling wind-throw in the Porcupine Mountains, Michigan, had parabolic patterns with continuously increasing growth for the first 25 years or less. The declining or parabolic pattern in the growth of initially dominant trees over several decades is well documented (Marshall 1927, Fritts 1976).

The last types of patterns to be mentioned here are the more ambiguous patterns that either increase slowly over many years or fluctuate irregularly over time (Figure 3.5). Radial growth patterns that increase gradually over many years present a difficult interpretive problem. The

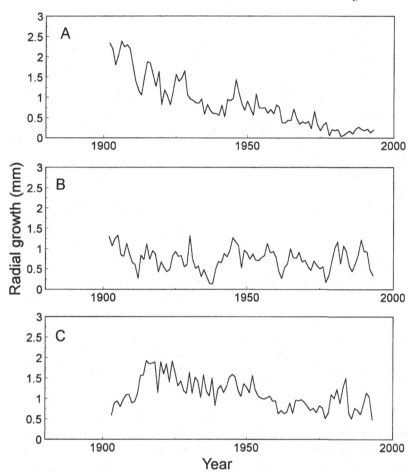

Figure 3.4. Examples of gradually declining (A), flat (B), and parabolic (C) radial increment patterns from the black spruce and jack pine forest studied by Frelich and Reich (1995a).

only thing one knows for sure about such a pattern is that if the tree was in the canopy at the time of coring then the tree must also have been in the canopy by the time the maximum ring-width on the core occurs. The tree could have been in the canopy all along, but in a poor competitive position, or there could have been a release followed by a drought or injury which has 'hidden' the release. The date of entrance into the canopy for such trees lies somewhere in an 'ambiguous zone' between the first ring and the time of maximum growth. Several solutions are possible, such as to pick the decade with the highest upward growth trend,

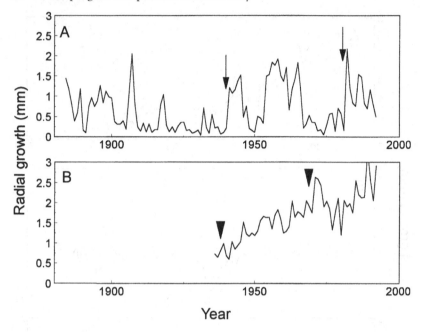

Figure 3.5. Examples of ambiguous and irregular growth patterns. A, releases indicated by pointers in a paper birch tree with irregular growth patterns and B, ambiguous zone delimited by triangles in a black spruce tree. Both trees were on the study area of Frelich and Reich (1995a).

or the middle of the ambiguous zone. If the earliest growth rates are high enough to indicate 90% or 95% chance that the tree was already in a gap (see analysis of early growth rates), then the tree can safely be assumed to be in the canopy from the start.

Irregular patterns are defined as those with two or more peaks 20 or more years apart, after smoothing the pattern with a 15 or 20 year running mean to exclude short spikes and valleys caused by climatic events. Irregular patterns could be caused by cycles of release and over-topping, by injury to the tree followed by gradual recovery or by unusual sensitivity to climate or site factors by an individual tree. Irregular patterns can be handled as special cases of the already mentioned patterns (e.g. Lorimer and Frelich 1989, Frelich and Lorimer 1991a). If the early growth rate meets the threshold value for early growth rate analysis, the tree can be classified as initially in a dominant position. If the height of successive peaks occurred in an overall flat or declining pattern and the highest part of the first peak is 25 years or less from the start, the tree can be classified as initially dominant, consistent with the interpretation of the parabolic pattern. If the highest part of the first peak is more than 25 years from the start or the height of successive peaks is increasing, there is an ambiguous zone. Of course, any of the peaks which are abrupt and sustained enough to meet the criteria for a release should be counted as such.

Summary of guidelines for interpretation of radial-growth patterns

Radial increment patterns are used to indicate when trees went from suppressed crown classes (i.e. overtopped by other trees) to intermediate, co-dominant or dominant positions (i.e. receive direct sunlight on the crown, and are in a canopy position). The following guidelines summarize my experience with radial growth patterns in the Great Lakes Region. These are general guidelines only; the reader may have to adjust them for use in other regions, special situations, or in light of new evidence.

I. Shade-intolerant tree species: assume that they were in the canopy from the start, regardless of the radial increment pattern. Analysis of total tree age is necessary to establish time of disturbance. Releases indicate stand-thinning events.

II. Mid-tolerant and tolerant species:

    A. Major release(s) present (including those in the first 15 years of the tree's ring record)
        • Tree in canopy gap at time(s) of release(s)

    B. Moderate release(s) present (including those in the first 15 years of the tree's ring record)

        1. Release is the first or only release on the tree's record
            • Tree in canopy gap at time of release

        2. Release is not first or only moderate release
            • A neighboring tree died at the time of release

    C. No releases present

        1. Growth pattern is flat, declining or parabolic
            • Tree was in a canopy position from the start of the ring record

        2. Growth pattern has slowly rising ambiguous zone
            a. Ring width exceeds threshold value for gap origin at sapling size class
                • Tree entered canopy gap when ring-width exceeded threshold

b. Ring width does not exceed threshold for gap origin at any size
- Tree entered canopy during decade of maximum upward slope within the ambiguous zone
3. Growth pattern is irregular
- Treat as a special case of any of the above conditions as appropriate.

Sampling stand age structure

Many studies of forest history require accurate reconstruction of forest age structure, and the choice of sampling method has a major impact on interpretation of results. Here I contrast two ways of viewing forest age structure, both of which are useful in certain contexts. Both cases assume that one has established study plots, and will either sample all trees or sample trees using a random or systematic method.

Tree-population-based age structure

Typically, age structure is expressed as a relative frequency distribution, and the proportion of individuals in each age class (or cohort) is:

$$RF_i = \frac{N_i}{N_t} \tag{3.3}$$

where:

RF_i is relative frequency of age class I

N_i is number of stems in age class I

N_t is total number of stems in the population.

Sampling could be a total population sample, where all trees on a plot are cored, or a random or systematic sample, representing a sub-set of all trees. If a sub-set of trees are to be cored, all trees are usually numbered and the sub-set chosen by number prior to coring in the field (e.g. Canham 1985).

Area-based age structure

The areal proportion of a plot or landscape occupied by each age class (or cohort) is:

$$RA_i = \frac{A_i}{A_t} \tag{3.4}$$

where:

RA_i is relative area of age class I

A_i is area occupied by age class I

A_t is total area occupied by trees on a plot or landscape of interest.

The sampling is carried out by selecting the tree stem or crown nearest to each of randomly or systematically located points on the ground. Area within a stand occupied by a given post-disturbance cohort can be expressed as the proportion of points, as represented by nearest tree stems, or in terms of proportion of crown area occupied by each cohort. Several studies have employed exposed crown area (ECA), defined as the horizontal area of the portion of a tree crown directly exposed to sunlight not including crown overlap of adjacent trees. The relative proportion of ECA for age class I would be the aggregate ECA of all trees in age class I, divided by the aggregate ECA of all trees in the stand. These two options will be discussed in more detail below.

Interpretation of the two sampling strategies
There are some major differences between tree-population-based and area-based age distributions. Population-based samples are often used by demographers to calculate life tables with growth and mortality rates. Mortality is expressed as percentage of individuals dying per unit time (reciprocal of average lifespan or canopy residence time). The shape of the tree-age distribution can be interpreted as even-aged or multi-aged. Both density-dependent mortality, or self-thinning, and density-inde-pendent mortality, or senescence and removal by disturbance, can be ana-lyzed. Because many small trees may occupy the same area as a few large trees, population-based samples give a poor estimate of areal extent of disturbance.

   Area-based age samples are commonly used in studies of disturbance regimes by foresters and landscape ecologists. Disturbance rate is expressed as percentage of land area disturbed per unit time (reciprocal of rotation period). Within stands, disturbance rate may be expressed as per-centage of canopy area (ECA) that turns over per unit time. Thus, area-based studies are useful for reconstruction of canopy turnover rates and gap dynamics within stands (e.g. Runkle 1982), and patch characteristics of a landscape (e.g. Heinselman 1973, Frelich and Lorimer 1991a, Frelich and Reich 1995b). Self-thinning cannot always be detected because a cohort may lose many individuals to density-dependent mortality but still continue to occupy the same area with fewer, but larger trees.

Tree-population- and area-based samples will yield the same result only in cases where the forest is strictly even aged, or if all trees are allocated the same amount of space regardless of age. One may visit a local apple orchard to see such a stand. However, many age-structure studies (other than those that sample all trees) do not make clear whether tree-population-based or area-based samples were used to choose sample trees, at least within some of the stands included in the study (e.g. Heinselman 1973, Romme and Knight 1981, Yarie 1981, Knowles and Grant 1983, Veblen 1986, Scott and Murphy 1987, Foster 1988ab, Platt et al. 1988, Stein 1988, Abrams and Scott 1989, Deal et al. 1991). Many of these papers state that trees were selected at random, or that selected trees were cored. However, whether the sampled trees were the nearest stem at random points on the ground or a random selection from the population is important for interpretation of the data. Of course, all of the cited studies make valid points, and all have useful information. But it is difficult for the reader to judge the overall quality of a study, or use the data for comparison with other studies, without knowing how sample trees were selected.

Compilation of stand disturbance chronologies

A disturbance chronology generally indicates what proportion of trees in a stand entered the canopy over time, commonly on a per decade basis in temperate forests. Now a simple example will demonstrate how to sample trees within a stand to make a disturbance chronology and also compare the properties of tree-population-based and area-based samples of forest age structure. I will also show how to convert one sample type to the other for applications where, for logistical reasons, the optimum type of sample cannot be used or both tree-population- and area-based samples of forest age structure are desired.

Comparison of sampling strategies

Differences between tree-population-based and area-based sampling methods are illustrated by sampling 25 trees (1/3) from a hypothetical 60 by 60 m mixed-age forest plot (Figure 3.6). A full tally of distribution of individuals, and distribution of ECA, among cohorts serves as the standard for comparison of three sampling strategies (Table 3.2). As many trials as possible of each sampling scheme were carried out and the

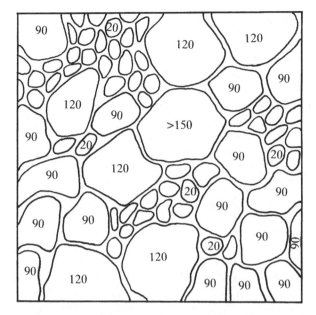

Figure 3.6. Tree-crown map (showing exposed crown areas) on a 60 by 60 m hypothetical forest plot. Numbers within tree crowns are age (years). Unlabeled small trees are part of the 20-year-old cohort.

average was used for comparison with the known values, because the point here is the overall comparison of the sampling schemes, rather than whether there are errors in an individual sample.

Strategy 1. Systematic by number – a tree-population-based sample. Trees were numbered consecutively and every third tree was sampled for a total of 25. Because 1/3 of the trees are sampled per trial only 3 trials are possible.

Strategy 2. Systematic by a grid of points – an area-based sample. A grid of 25 points was laid over the map in Figure 3.6 and the tree with crown overhead (or nearest crown) was sampled at each point. Ten trials were carried out by shifting the origin of the grid. This represents the maximum possible number of trials, since no additional shifts of the grid could be done without obtaining a set of trees identical to a previous trial.

Strategy 3. Stratified random – a combination of area- and- population-based samples. The plot was divided into 25 grid boxes and one tree was picked randomly by number within each box. Five trials were obtained without excessive duplication.

Table 3.2. *Comparison of stem number and exposed crown area for the hypothetical forest plot*

	Cohort age (yr)			
	>150	120	90	20
Number of stems	1	6	17	51
Stems (%)	1.3	8.0	22.7	68.0
ECA (m²)	228	1026	1276	611
ECA (%)	7.3	32.7	40.6	19.4

Notes:
Only canopy trees (receiving direct sunlight on the crown) are considered in this example.
ECA, exposed crown area.

The example (Figure 3.6) was purposely devised to show the major difference between tree-population-based and area-based measures of dominance within closed-canopy forests. For example, the 90-year-old cohort has only 22.7% of the stems, but occupies twice as much area as the 20-year-old cohort with 68% of the stems (Table 3.2). Samples taken using strategy 1 (tree-population-based) agree closely with the known relative-frequency versus age distribution of stems (cf. Table 3.2, Table 3.3). Sampling strategy 2 (area based) agrees with the known relative area versus age distribution, or area occupied by each cohort (cf. Table 3.2, Table 3.3). The stratified random sample does not gauge either percentage of stems, or percentage of area occupied, but seems to be intermediate between the two (cf. Table 3.2, Table 3.3).

Extreme care should be used before deciding on any type of spatial stratification of age samples within stands. Stratified samples may not measure either age class–frequency or age class–area distributions, and it is questionable what the meaning of the data is. This problem would be minimized in relatively uniform forests and maximized in forests with complex age structure with overlapping spatial distribution of age classes. Any sampling strategy that may allow trees in different locations relative to the grid corner or grid center to be selected within a given grid cell is a stratified sample (e.g. numbering trees within each grid cell and then selecting one at random). If the rule within each grid cell was to select the tree nearest the center the resulting sampling strategy would be systematic (i.e. would give the same result as strategy 1) rather than stratified.

Table 3.3. *Estimated age distributions (relative frequency or relative area) from sampling schemes 1, 2 and 3*

	Cohort age (yr)			
Sampling strategy	>150	120	90	20
(1) Tree population based		% stems		
Mean	1.3	8.0	22.7	68.0
Range	0.0–4.0	8.0–8.0	20.0–24.0	68.0–68.0
(2) Area based		% ECA		
Mean	7.2	31.2	40.8	20.8
Range	4.0–8.0	24.0–36.0	36.0–44.0	16.0–24.0
(3) Stratified		?		
Mean	0.8	20.0	28.8	50.4
Range	0.0–4.0	16.0–24.0	20.0–40.0	48.0–60.0

Although it is generally clear whether a given study requires population-based or area-based sampling, logistical considerations may ultimately determine which strategy is chosen. For example, in studies of population dynamics over large areas, numbering of trees necessary for a population-based sample may be impossible. In this case, if area-based sampling is carried out, data must also be obtained on relative sizes of trees in different age classes for the purpose of converting the data to relative frequencies of stems.

It is perfectly feasible to do population-based samples, and then convert to area-based data, so that both life tables and canopy disturbance rates can be analyzed in one study. One caveat is that small sample sizes of large, old trees will often result. This is especially true in very old stands where there may be tens or hundreds of small trees for each large tree. In this case additional large trees may be sampled to look at their age structure separately. When an age distribution is compiled for all age classes in the stand the representation of large trees in the sample can be weighted proportionally lower (e.g. if their representation in the sample is twice that in the stand, weight their data by half).

Conversion between tree-population and area samples
Conversion of tree-population-based samples to area-based estimates can be done if exposed crown area (ECA) of all sample trees is available.

Exposed crown radius can be measured relatively quickly in the field while taking the increment core of each tree so that ECA can be calculated in the lab (Lorimer and Frelich 1989). Alternatively, the ECA for each tree may be predicted from an ECA versus dbh regression. In either case the ECA is tallied for all sample trees within each cohort or age class and then expressed as a percentage of the sum of ECA for all sample trees:

$$RA_i = \frac{\sum_{j=1}^{N_{si}} ECA_{si}}{\sum_{j=1}^{N_{st}} ECA_s} \tag{3.5}$$

where:

RA_i is estimated relative area of age class I

j indexes sample trees

N_{si} is the number of sample trees in age class I

N_{st} is the total number of sample trees

ECA_{si} is the exposed crown area of a sample tree in age class I

ECA_s is the exposed crown area of a sample tree.

Conversion of area-based samples to tree-population-based estimates is made by dividing the proportion of total area occupied by each age class by the average ECA among sample trees for each age class. The result is a 'pseudo proportion' in each age class. When this is done for all age classes the resulting 'pseudo proportions' are rescaled to add to 1:

$$RF_i = \frac{PCA_{si}}{ECA_{si}} \frac{1}{\sum_{j=1}^{N_a} PCA_{si}/ECA_{si}} \tag{3.6}$$

Where:

RF_i is estimated relative frequency of age class I

j indexes sample trees

N_a is the number of age classes or cohorts

ECA_{si} is average exposed crown area of sample trees in age class I

PCA_{si} is proportion of sample trees in age class I.

If we apply these conversion formulas to the example we see that the interconverted estimates of proportion of trees and area occupied by trees in each class are close to the actual values (cf. Table 3.4, 3.2).

Table 3.4. *Conversions from tree population to area occupied samples and vice versa*

Conversion type	Cohort age (yr)			
	>150	120	90	20
Tree population to area occupied	% ECA			
Mean	6.6	32.8	41.0	19.6
Range	0.0–19.8	29.8–34.3	32.7–45.2	17.7–20.5
Tree population to area occupied >150 fixed at 7.3%	% ECA			
Mean	7.3	32.7	40.5	19.5
Range	–	31.8–34.4	37.8–41.9	19.0–20.5
Area occupied to tree population	% stems			
Mean	1.4	7.2	22.5	69.0
Range	1.3–1.4	7.2–7.3	19.5–28.3	63.1–72.0

The sampling strategies and formulas for conversion given here apply mainly to canopy trees in dense forests. They also assume that only one canopy layer is being analyzed. Therefore, understory trees have to be handled separately. It is not known how the area-based sampling strategy would respond when applied to savannas with discontinuous tree cover. Conversions between population-based data and area-based data should be used with caution in savannas, because trees in groves may have different crown sizes for a given dbh than trees growing singly on the savanna. The example given here highlights the necessity for researchers to work through examples such as shown in Figure 3.6 before starting a major forest-disturbance study, to see if the proposed sampling scheme/number of increment cores will work given the spatial mix of cohorts in a specific study area.

How to handle trees without complete cores

In humid climates, many older trees are hollow, and it is often impossible to determine the date of canopy accession. Sometimes large tree size also makes it physically impossible to get complete cores from some trees in each stand. Such trees must, however, be included in any sampling scheme. If they are not included then there is great risk that younger cohorts will look more important than they truly are. The ECA of such

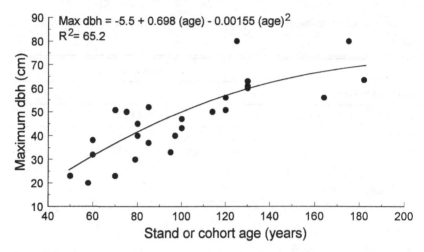

Figure 3.7. Maximum dbh versus age relationship for 25 sugar maple stands in western Upper Michigan. The regression line indicates a conservative minimum length of stand-chronology reconstruction. For example, if good tree core data come from trees 40 cm dbh or less, then 80 years is the maximum length of time that a chronology could be carried back in time, while data from trees up to 60 cm dbh would allow reconstruction back 130 years.

trees can still be measured, so that the proportion of ECA occupied by all other cohorts is known. For example, the >150-year-old tree in Figure 3.6 may be hollow, but we would still know that it occupies 7.3% of ECA. In the example the samples from the other trees can be weighted so that they add up to 92.7% (Table 3.4). It is simply wrong to avoid the old hollow trees by always selecting a new tree every time a hollow one is encountered, and then come to the conclusion that younger cohorts occupy 100% of the stand.

There are two ways to handle these hollow trees. One is to include them in the sampling, recording species, dbh, and other relevant data, and continue to select new trees until the desired number of complete cores is in hand. Then the proportion of incomplete cores and the size of the trees they are from is known. The second way to handle hollow trees is to determine a threshold diameter above which the proportion of hollow trees is so high that it is not worth coring trees above the threshold. One can then limit the length of the disturbance chronology for the plot to the minimum time it would reasonably take trees to reach that threshold size, by following the upper 5th or 10th percentile of an age–diameter relationship (Figure 3.7). Coring could then be concentrated among size classes where nearly all trees are sound, resulting in a more accurate disturbance chronology for a shorter time period.

Those studying forest disturbance history also should not forget that recent gaps, where trees are too small to core, must also be taken into account. Trees in these gaps are part of the forest canopy in that they receive direct skylight on top of their crowns and they are in the process of beginning to respond to a canopy disturbance. Few people take cores from trees less than 5 cm dbh. To obtain a complete sample of all canopy accession dates one must record the number of sample points that fall within gaps where trees are too small to core and somehow estimate the ages of all such gaps. For a complete chronology up to the time of sampling, one must measure these recent gaps, cut small saplings down to estimate gap age, or else consider the gaps as an unsampled part of the forest.

Regardless of how trees that cannot be cored are handled in the field, it will be necessary to construct an ECA–dbh regression to estimate the proportion of ECA occupied by these unsampled trees. Frelich and Lorimer (1991a) predicted the ECA of large hollow trees (>60 cm dbh) that often occupied 10–30% of total ECA on their 70 plots in Upper Michigan. Only species and dbh need be recorded for large unsampled trees if an ECA–DBH regression is available, making field work efficient. The regression approach to ECA also may have the advantage of being useful on a regional basis (Lorimer and Frelich 1989).

Once the proportion of ECA not sampled is known, the true proportion of the stand represented by each cored tree (WPCA) can be calculated:

$$WPCA = \frac{W(100)}{N} \tag{3.7}$$

where:

WPCA is the weighted percentage crown area

W is the weighting factor

N is the total number of cored trees on the plot.

The weighting factor, W, is calculated as:

$$W = \frac{ECA_s}{ECA_t} \tag{3.8}$$

where:

ECA_s is the aggregate exposed crown area of all trees in size classes that were sampled

ECA_t is the aggregate exposed crown area of all trees on the plot.

For cases where sample trees were picked on a stratified scheme by tree size (e.g. core the nearest pole-sized, mature and large tree at each grid point), a more complicated method for calculating W is necessary that takes into account the ECA in each tree size class:

$$W = \frac{(ECA_d/ECA_t)}{Y/N} \tag{3.9}$$

where:

W once again is the weighting factor

ECA_d is the aggregate exposed crown area occupied by trees in diameter class d, from which the core was taken

ECA_t is the aggregate exposed crown area of all trees on the plot

Y is the number of cores in diameter class d

N is the total number of cored trees on the plot.

Note that when trees are chosen without regard to size class and there are no problems with hollow trees, so that all size classes are sampled, the weighting factor becomes 1.0, and ceases to become a factor in the analysis. Each core then represents $1/N$ of the forest canopy, where N is the number of sampled trees.

Handling changes in cohort area over time

Disturbance chronologies as described above represent the area occupied by various age classes, currently or in the past, with different degrees of accuracy. Two factors that determine accuracy are: (1) the tendency of trees from the original cohort to die over time so that the cohort occupies a smaller amount of space as time passes; and (2) the opposing tendency of pole and mature sized trees to expand their crowns over time so that the cohort occupies more space over time.

How fast does cohort area shrink over time?

Self-thinning occurs in even-aged cohorts so that fewer but larger trees occupy approximately the same area as time goes on. After a certain age, however, density-dependent mortality ends and additional mortality reduces the area occupied by an age class. Eventually gaps formed by the death of a tree will be so large that crown expansion of other trees in the cohort can no longer compensate for a dead individual. At this point the relationship between area occupied by an age-class and size of the original disturbance breaks down. This age can be estimated for a typical stand in any forest type. For example, in the hemlock–hardwood forest, Runkle (1982) found that the average dbh of a gapmaker was about 50 cm, while the minimum gapmaker dbh (2 of 2921 trees) was about 25 cm. Runkle estimated that the average age at the time a tree dies and forms a gap ranges from 160 to 255 years for sugar maple and from 182 to 366 years for hemlock. The ranges are derived from different studies of age–diameter relationships throughout the northern hardwood forest. However, sugar maple and hemlock will reach the minimum gapmaker dbh (25 cm) in virgin stands after 130 to 190 years on the average (Gates and Nichols 1930, Morey 1936, Tubbs 1977). Thus, the possibility that mortality would significantly reduce the area occupied by an age class in this forest type could be ruled out by using the average age for trees to attain the minimum gapmaker size, 130 years in this example, as the maximum length of time that the disturbance chronology adequately represents the original cohort area.

Crown expansion

Now, let's deal with the second issue of crown expansion by trees in a cohort. Accurate estimates of the areal extent of disturbances that occurred decades ago depends on the assumption that the area occupied by a disturbance age class remains approximately constant within the time frame of self-thinning discussed in the previous section. However, mature trees adjacent to newly formed gaps usually expand their crowns into the gap. Thus, some years later the area occupied by the cohort in the gap may appear to be smaller than it was at the time of gap formation, and the older cohorts around the gap may appear larger.

To become established in the canopy a newly released sapling in a gap must grow at least as high as the widest part of the crowns of surrounding canopy trees without being overtopped. The questions to be answered are therefore:

1. How high is the widest part of the surrounding tree crowns?
2. How high are typical suppressed saplings at the time of release?
3. How long will it take the saplings to make up the difference from their height to the height of the surrounding trees?
4. How much horizontal crown expansion will the surrounding trees accomplish during the time (from question 3) that the saplings are growing up into the gap?

Table 3.5. *Estimated crown expansion by sugar maple gap-border trees that will occur before recently released trees in a new gap are established canopy members*

	Initial dbh – border trees[a]			
	5.5	18.5	36.0	56.0
Height (y) of widest part of crown (m) of border trees	6.6	10.1	12.8	14.9
Years (x) for gap tree to grow from 5.0 m to y	6	18	28	35
Dbh (cm) of gap tree at x years after gap formation	3.1	7.8	11.4	13.7
Crown radius (m) of gap tree at year x	1.6	1.9	2.1	2.3
Final dbh (cm) of border trees at year x	6.3	26.9	44.9	67.2
Estimated crown expansion of border trees (m)	0.6	2.3	2.1	1.6

Notes:
[a] The four initial dbh are the means for the sapling (0–10.9 cm dbh), pole (11.0–25.9 cm dbh), mature (26.0–45.9 cm dbh), and large (≥46.0 cm dbh) trees.
Source: After Frelich and Martin (1988).

These questions can be answered by intensive analysis of tree growth data. For sugar maple–hemlock forest in Upper Michigan, Frelich and Martin (1988) found that a typical suppressed sapling was 5.0 m tall at the time of release. Such a sapling would have to grow another 1.6–9.4 m to reach a point where it was no longer in danger of being overtopped – that point where it was as tall as the widest part of the crowns of trees surrounding the gap – depending on the size of the trees surrounding the edge of the gap. Gap edge trees ranged from saplings at 5.5 cm dbh through pole, mature and large trees with a mean dbh of 56.0 cm (Table 3.5). Age versus height regression showed that a typical sapling would take 6 to 35 years to grow the required distance and that the surrounding trees would expand from 0.6 m to 2.3 m into the gap, depending on their size class, during that time (Frelich and Martin 1988, Table 3.5).

At first thought, it seems that this expansion by trees bordering gaps would inherently underestimate the relative area occupied by gaps, especially if the forest is sampled years after an episode of gap formation. However, it turns out that this actually depends on the method by which sample trees used to reconstruct stand history were selected. At each sample point, one has the choice of coring the tree with the nearest stem, or the tree with the crown overhead (or nearest crown overhead if the sample point happens to land in the interstitial space created by crown shyness).

For recently formed gaps, border trees will nearly always have a crown radius larger than new saplings inside the gap. Therefore, sample points which land outside the gap, under the crown of an adjacent large tree,

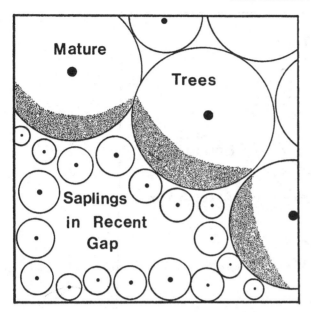

Figure 3.8. Edge of recent gap filled with saplings surrounded by mature trees, in aerial view. Points on the ground beneath the shaded zones are closer to one of the saplings than to the trunk of the mature tree whose crown is overhead. From Frelich (1986).

may still be closer to the stem of one of the new saplings inside the gap (Figure 3.8). The average effect of this error is to make the radius of the gap appear larger than it actually is, so that the proportion of cored trees that are estimated to be inside gaps is larger than the proportion of the plot occupied by the gaps. Thus, the nearest-stem method of selection would overestimate the extent of recent disturbance, while the overhead-crown selection method would have no error.

The magnitude of this type of sampling error, and the relative performance of nearest-stem versus overhead-crown selection methods, will depend on the relative sizes of border trees and the new trees inside the gap, which will change over time as border trees expand into the gap and new trees grow. Let us examine the situation 35 years after gap formation in a hemlock–hardwood forest, when, according to the above discussion, gaps trees will have had time to become established as canopy members. At this point crown expansion stops unless additional new gaps have formed. An average mature tree (36.0 cm dbh) in this forest on the edge of a new gap has a crown radius of 3.6 m, and will expand into the gap by 1.9 m – for a total radius of 5.5 m – by the time the gap saplings have

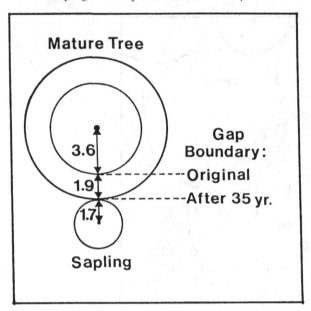

Figure 3.9. Hypothetical example of expansion of mature trees into a gap. Distances from the 35-year-old sapling and the mature tree to the original gap boundary are equal in this instance. From Frelich (1986).

become established canopy members and crown expansion stops (Frelich and Martin 1988). Since an average sapling has a radius of 1.7 m, the total distance between the mature border tree and the sapling in the gap will be about 7.2 m (Figure 3.9). The midpoint of this distance (3.6 m) falls exactly at the original boundary of the gap (Figure 3.9). Therefore, the nearest-stem selection method would accurately estimate the areal extent of the gap at the time it formed, but it would also underestimate the area currently occupied by the mature trees that were around the border of the gap when it formed. The overhead-crown selection method, in complementary fashion, would underestimate the original extent of the gap, but correctly estimate the current extent of occupancy by the expanded gap-border trees.

We can adapt this analysis to examine all gap and border tree size classes, using the appropriate crown expansion estimates for each size class, and calculate the average sampling error for a large stand with many gaps (Table 3.6). To combine the errors in Table 3.6 into an average for all size classes, the average proportion of gap circumference bordered by each size class must be calculated. Since trees bordering a gap are tangent

Table 3.6. *Apparent error in gap radius caused by crown expansion for two methods of selecting sample trees*

	Initial size of border trees				
	Sapling	Pole	Mature	Large	
Proportion of average gap circumference occupied	0.26	0.22	0.26	0.26	
Selection method	Average error (m)				Weighted average
Nearest stem	−0.2	−0.8	−0.2	0.6	−0.1
Overhead crown	−0.6	−2.3	−2.1	−1.6	−1.6

Source: After Frelich and Martin (1988).

to the gap (meaning they do not by definition intersect the gap border) the average proportion of gap circumference bordered by a size class is not the same as the proportion of the total crown area occupied by the size class. The actual proportion is related to the relative number and crown diameter of trees among the size classes. For example, suppose that there are two size classes, large and small, in a forest. Suppose that the large trees occupy 75% of the forest and have an average crown diameter and crown area of 12 m and 113.1 m². Suppose that the small trees occupy 25% of the forest and have an average crown diameter and crown area of 4 m and 12.6 m². The ratio of number of small to large trees is:

$$\frac{25\%}{12.6 \text{ m}^2} : \frac{75\%}{113.1 \text{ m}^2} \text{ or } 1.98\%/\text{m}^2 : 0.66\%/\text{m}^2$$

This can be simplified to three to one. If trees are selected at random and placed along the border of a gap, three of the small trees with a crown diameter of 4 m will be placed for every large tree with a crown diameter of 12 m (Figure 3.10). If crown diameter is used as an estimate of the portion of the gap border taken up by each tree, it is clear that the two size classes will occupy equal proportions of the border. The procedure just discussed can be expressed mathematically as:

$$P_i = \frac{(D_i)(N_i)}{\sum_{I=1}^{n} (D_i)(N_i)} \tag{3.10}$$

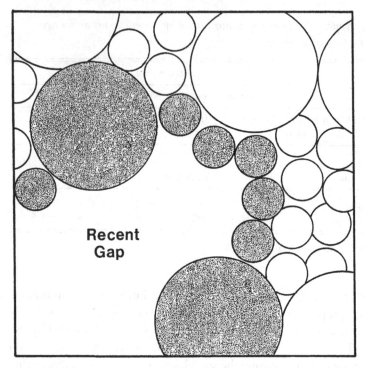

Recent
Gap

Figure 3.10. Proportions of gap boundary occupied by small and large trees. In this example, three small tree (4 m crown diameter) are present for each large tree (12 m crown diameter). From Frelich (1986).

where:

P_i is the proportion of circumference of an average gap bordered by dbh size class I

D_i is the average crown diameter of dbh size class I

N_i is the relative number of trees of dbh size class I

n is the number of dbh size classes.

Average proportions of gap circumference occupied by sapling, pole, mature and large trees in the hemlock–hardwood forest was calculated using this formula, using pooled data from a large number of plots (Frelich and Martin 1988). The results (Table 3.6) show that the average proportions of gap circumference occupied by the four size classes, sapling, pole, mature, and large, in this case are fairly uniform. The average error for both selection methods, for gaps that have been closed by gap-sapling growth into the canopy, can be weighted by the relative abundance of trees in different size classes to obtain an overall stand-wide error. If one is interested in examining the original area of gaps at the time they were formed, the nearest-stem selection method provides very little error in this forest (only -0.1 m on average), while the overhead-crown selection method provides for an average error of -1.6 m (Table 3.6). Since gap perimeter is very large in old multi-aged forests, an average error of 1.6 m is very significant, although it would not matter much when sampling young even-aged stands.

To summarize this crown expansion factor directly, we may say that there is an important difference between selecting the tree with crown overhead versus the tree with the nearest stem at points on the ground. Because crowns of pole and mature trees expand aggressively, the areal extent of crowns of a cohort may be larger than the original gap in which the cohort was recruited. The stems of a cohort in a gap do not very well define a recently formed gap, but may define the area of the original gap after a few decades and continue to do so for many more decades. Therefore, the overhead-crown selection method should be used when the goal is to estimate the current area occupied by a cohort regardless of its date of origin, but the nearest-stem selection method should be used for a better estimate of original extent of older cohorts that have closed gaps (Frelich and Martin 1988), keeping in mind the length of time limitation discussed above (see the previous sub-section of this chapter: 'How fast does cohort area shrink over time?'). For recently formed gaps, the area occupied by the crowns of the young saplings and the area of the original gap are the same thing. Therefore, if one's goal is to reconstruct canopy turnover rates, which require knowledge of the actual area of gaps, a hybrid selection method may be in order: if a sample point falls in a gap, get the age (or core if possible) of a gap tree, but if it falls in a closed gap (*sensu* Runkle 1981, 1982), core the tree with the nearest stem.

How many trees to core?

If one will not or cannot sample all trees, then there is always a question of how many trees to sample. The main considerations are the chance of missing a cohort of trees in complex multi-aged stands, and the precision of the estimated proportion of trees in each cohort.

The probability of failure to detect an age class of trees can be calculated for any size of cohort and number of sample points:

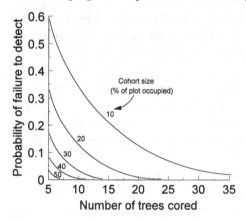

Figure 3.11. Probability of failure to detect a cohort for *x* number of randomly located cores within a forested plot of trees.

$$Pf = (1 - Py)^X \tag{3.11}$$

where:

Pf is the probability of failure to detect age class *y*

Py is the proportional area occupied by age class *y*

X is the number of independent sample points.

The chance of missing a cohort that occupies 50% or more of stand area falls to less than 5% with as few as five trees cored (Figure 3.11). To achieve this level of detection in complex stands with many cohorts that each occupy a small proportion of the stands requires at least 30 cored trees, randomly or systematically distributed throughout the stand.

The precision of estimates of the proportion of a stand occupied by each cohort can be given by the confidence limits for proportions:

$$p + 1.96 \sqrt{\frac{p\,(1-p)}{N}} \tag{3.12}$$

where:

p is the sample estimate of proportion of points belonging to a given cohort

N is the number of sample points (assumed to be independent with respect to date of canopy accession).

Independence of sample trees

All parametric and non-parametric statistical confidence intervals and tests require independent samples. Contrary to most scientists' understanding, neither a so-called random sample nor a sampling scheme where every location has an equal chance of being sampled guarantees independent samples. A systematic sample does not rule out statistical analyses in the context of forest disturbance. Many have assumed that choosing samples (in this case trees to core) at random is equivalent to choosing independent samples. However, disturbance is a spatial process, and randomly chosen trees do not guarantee independence. For example, in cases where a stand comprises a few large patches, each originating after a different disturbance, all cores taken from trees in each age class are not independent, regardless of whether they were done in a systematic grid or randomly. Therefore, the disturbance chronology is merely an empirical observation of approximate proportions in each age class, and no statistical test can be done. This information may still be very valuable, however.

One can find out if tree ages are independent within a stand, but not until the stand chronology is finished! Then, if the cored trees have been mapped, one can check for spatial autocorrelation, and if that falls to an insignificant level at some distance – say 20 m – then any trees sampled that are 20 m or more apart are 'independent' for purposes of statistical tests. This criterion also applies to cases where the objective is to see if disturbance rates are significantly different on different slopes or other landforms within a study area. Moran's I may be used to check for contagion in tree ages (Sokal and Oden 1978, Frelich and Graumlich 1994) and is also available in a two-dimensional form (Czaplewski and Reich 1993). Contagion among categorical variables, such as tree species or age classes, can be assessed by using Ripley's K, the standard normal deviate method, or other spatial statistics (Sokal and Oden 1978, Legendre and Fortin 1989).

For all practical purposes, a systematic sample works best for disturbance history analyses. They are more accurate at assessing the areal extent of each disturbance event than random samples (Payandeh and Ek 1971, Snedecor and Cochran 1980), which often have many samples clustered in one area, and leave other parts of the stands undersampled. Also, systematic samples are much easier to carry out in the field, as anyone can tell you who has ever tried to locate a bunch of random points in a dense forest, where it is often necessary to measure from some point of origin to locate each point. If random samples are logistically difficult to establish, are not as accurate as systematic samples, and do not guarantee independence of samples, there is no point in using them.

The stand structural chronology

A second type of chronology can be constructed from tree-ring data that illuminates the effects of disturbance on the structure of stands. With the conventional chronology discussed above, the question to be answered is how many individuals or how much area is in each cohort? The alternative question is: how much of the stand area was occupied by saplings (0–10.9 cm dbh) or saplings plus poles (11.0–24.9 cm dbh)? Then instead of looking for evidence that most trees entered the canopy at the same time, one can answer other questions such as: did the stand ever experience series of disturbances such that the structure was reduced to that of a sapling or pole stand, even though there was never a stand-replacing disturbance? In some ways, this analysis is more important than the standard disturbance chronology. Wildlife populations, for example, respond more to the size structure of stands than to age structure.

Thus, a chronology can be constructed that shows proportion of the stand occupied by saplings (trees <10.9 cm dbh) or trees in any other size class of interest in each decade. The same field sampling and calculations described for the standard disturbance chronology are needed. For the sapling anaysis just mentioned the weighted percentage crown area for each sample tree is entered on the chronology for all decades from the earliest rings until the tree exceeds the sapling upper diameter limit. Then the weighted percentage crown area for all trees below the threshold size is summed to estimate the proportion of the plot occupied by saplings in a given decade (Figure 3.12).

The fraction of a plot occupied by saplings may represent the effects of more than one disturbance. For example, suppose that two disturbances, each of which removes 40% of the canopy of a stand, occur 20 years apart. Because it takes about 30 years for hardwood trees in the Lake States to grow to 10.9 cm dbh, the stand will be almost totally reduced to saplings after the second disturbance. Such episodes of disturbance have been referred to as 'functional catastrophes' (e.g. Frelich and Lorimer 1991a,b).

The approximate trunk diameter of a tree may be reconstructed for any point in the tree's past, assuming all ring widths were measured and the rings are satisfactorily cross-dated. To do this, one can simply add up

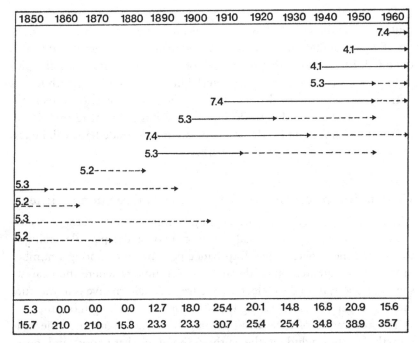

1850	1860	1870	1880	1890	1900	1910	1920	1930	1940	1950	1960

Figure 3.12. Example of a stand structural chronology from a nearly steady-state sugar maple–hemlock stand in the Porcupine Mountains, Michigan. The weighted percent crown area of each sample tree was entered on the chronology for each decade from the time of first canopy accession, until the tree reached the upper limit of the sapling size class (10.9 cm dbh, shown by solid line) and the upper limit of the pole size class (25.9 cm dbh, shown by dotted line). Box at the bottom shows the sum of weighted percent crown area for all saplings in the upper row, and all sapling plus poles in the lower row, for each decade. From Frelich (1986).

the cumulative sum of ring widths from the center to the desired year. The main problems with this technique are the bark (tree diameter in the field is usually measured outside the bark), and trunk asymmetry. If two or more cores are taken per tree, then the asymmetry problem is reduced, because one can use the average ring width for each year. However, almost all ecologists studying disturbance history would rather take one core per tree and core more trees, thus obtaining a more extensive disturbance history. Therefore a correction factor (CF) should be used when calculating tree size in the past: CF is the ratio of the radius of the sample increment core to the radius of the tree inside bark, as measured in the field. Some knowledge of tree bark thickness and the assumption that the cored radius maintained the same relationship to the tree's radius throughout the ring record are required. When this correction factor is multiplied by the core radius up to a given year and then a bark correction factor is added on, a much better estimate of tree size in that year is obtained. I have found that the length of some cores containing the pith is only one-quarter the tree's diameter! This occurs in old-growth forests with highly asymmetrical trees. This correction factor also corrects for shrinkage of the core due to drying. One wants to estimate tree size in live, wet condition. Thus, the average core radius inside bark will be less than half the tree diameter, inside bark.

Calculation of canopy turnover rates and residence times

The most direct method for calculating canopy turnover rates (percentage turned over per year – usually at the stand or landscape scale) is simply to take the mean percentage disturbance per unit time among a number of plots, from an area-based disturbance chronology where the nearest-stem method was used to select sample trees. An alternative is to measure the area of gap formation along transects for a known length of time.

The technique of measuring gap area does not work in some old-growth forests, including the author's hemlock–hardwood and near-boreal jack pine, spruce–fir forests. The reason is that there are few discrete gaps in these forests. The whole forest is laced with a network of interconnected small openings. When one tries the methods of Runkle (1981, 1982) in these forests, one can walk for long distances without coming to a well-defined gap edge. Instead of measuring regeneration in 'gaps' in these forests, one simply finds individual gap-making trees within a transect 10–20 m wide, and looks at the trees most likely to replace each gapmaker. The replacement species can be selected in different ways: the species with the highest number of stems within a few meters of the gapmaker, or the species with the biggest stem in the gap, or the species of the one stem that looks to the investigator to be the most logical successor to the downed tree. A species-by-species transition matrix is constructed directly from the raw data (e.g. Frelich and Reich 1995a).

The canopy residence times for individual trees, or the time from canopy accession until death, is the reciprocal of the turnover rates. Analyses may be done for all species pooled, or separated for each tree species or group of species (e.g. Runkle 1981, 1982). A more direct method of calculating canopy residence time is to take cores or wedges from trees that recently died and formed gaps. These cores are subjected to radial increment pattern analysis, and the number of years from release, rapid early growth, or start of a parabolic, flat or declining growth pattern until death, is the residence time for an individual tree.

Note that canopy residence times from one or two stands, or from many stands over a short period of time, may have little meaning. Ten-year canopy mortality rates within one stand are often very low; inspection of 70 plot chronologies from upper Michigan (Frelich 1986), shows that no mortality was detected for a total of 343 out of 742 decades of total reconstruction, or 47% of all decades. Also, average mortality per decade ranged from 1.7% to 18.9% among the 70 plots over the 120-year time window that was reconstructed.

To test the variability in estimated disturbance rates for different periods of observation, Frelich and Graumlich (1994) calculated the range of variability in estimated canopy residence times among observation periods from 10 to 60 years from an intensively studied 5-ha multi-aged hemlock–hardwood forest (Figure 3.13). Results show that at least 60 years of observation are necessary to get relatively stable estimates of canopy turnover within one stand (e.g. 170–200 years). Runkle's (1982) data cited above included only 10 years, but the data were taken from a large geographical area. Frelich and Lorimer's (1991a) data were based on

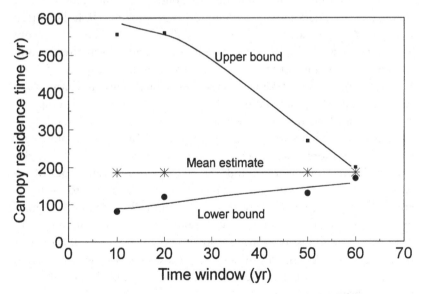

Figure 3.13. Range in estimates of canopy turnover for different lengths of observation or reconstruction, for a sugar maple–hemlock forest in Sylvania Wilderness Area, Upper Michigan. Data from Frelich and Graumlich (1994).

a large geographical area and long time window (120 years). Thus, it is clear that either a long time or large area, or both, is necessary for accurate estimates of canopy disturbance rates in the northern hardwood forests.

Summary

There are several lines of evidence for interpreting stand disturbance history. Each has problems; however, interpretation of radial increment patterns seems to yield the most accurate information in temperate forests. The proportion of trees entering the canopy (i.e. went from shaded understory position to a gap position with direct syklight on the crown) can be estimated by coring a number of trees and analyzing their radial increment patterns. As a general guideline, major releases (abrupt, sustained, 100% increase in ring width) indicate the time of canopy accession. Trees with parabolic arches near the center of the core, or flat or slowly declining growth patterns over the life of the tree were likely in a gap from the time of the first ring on their core. A disturbance chronology can be assembled by calculating the proportion of trees that entered the canopy in each decade.

Analyses of forest history require that hollow trees and trees too small to be cored be taken into account during sampling. If hollow trees occupy 30% of the area, then the area of all other trees can only occupy 70%, even though 100% of the sample cores come from them. Some studies of disturbance history have failed to account for unsampled trees.

Great care must be taken to determine which trees, and how many trees, to core. The following guidelines apply in cool-to-cold-temperate forest where I have the most eperience. In simple even-aged stands 5–10 cores may be sufficient to estimate stand history. In complex multi-aged stands 30 or more cores may be necessary to detect all cohorts that occupy 5% or more of the stand. There are two different types of age structure: the proportion of stems in each age class and the proportion of area occupied by each age class. It is important to know which is appropriate for a given study, because the sampling stategies are different. One has a hard time to justify calling a disturbance a 'catastrophe' if 80% of all trees originated after the disturbance, but those 80% of trees occupy only 10% of total crown area. There is much confusion in the literature on these methods and objectives.

To estimate the proportion of trees in each age class, all trees should be numbered and the sample trees selected at random from all individuals.

To estimate proportion of area occupied by various cohorts one can select trees that are closest to randomly or systematically located points on the ground. The proportion of area occupied by cohorts can also be reconstructed two ways: the area currently occupied by each cohort, or the area originally occupied by each cohort at the time it was a new gap. Sampling the tree with crown overhead at each sampling point on the ground will allow straightforward reconstruction of the area currently occupied by each cohort in a stand. In contrast, sampling the tree with the nearest stem at each sampling point will allow reconstruction of the area originally occupied by each cohort for about 120 years. Up to age 120, crown expansion by new trees in a gap approximately balances the loss of crown area within the gap by tree death due to self-thinning and the loss of gap area to expansion of trees surrounding the gap. Beyond 120 years these balances no longer hold and it becomes difficult to estimate the area originally occupied by a gap.

4 · *Disturbance, stand development, and successional trajectories*

Introduction

What are the major changes in stand structure over time (stand development) and what changes in composition over time (succession) accompany developmental changes? How do these changes occur in major forest types of the Lake States forests? These are the major questions examined in this chapter. Many studies on stand dynamics in the Great Lakes forests – studies which used methods described in Chapter 3 – are synthesized here. The chapter starts with basic schemes for development and succession. Then I proceed to case studies of five major forest types that illustrate the types of variations on the basic schemes of development and succession that occur among forests.

Basic stand development sequence

Four basic stages of development that stands go through after stand-replacing disturbance were described by Oliver (1981). I present these stages in a form modified to fit the context of cold-temperate hardwood–boreal transition zone forests of the Lake States (Figure 4.1).

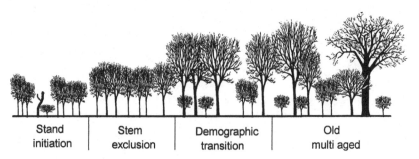

| Stand initiation | Stem exclusion | Demographic transition | Old multi aged |

Figure 4.1. Four basic stages of stand development.

Complications of this basic scheme of four stages that occur in different forest types will be described in detail later on this chapter. Stages of development are important to keep in mind when thinking about disturbance regimes throughout the rest of the book because it is disturbance regimes that determine the proportion of stands in each stage of development. For example, with short-rotation logging, all stands across the landscape will be in the initiation and stem-exclusion stages of development with no chance to go on to the old multi-aged stage.

Stage 1. Initiation. *This stage follows major disturbance, such as stand-leveling wind, crown fire or clear-cut logging.*

- The open space is filled in with individuals that arrive by seed (e.g. paper birch and aspen after fire), stump sprouts (e.g. oak forest after fire), roots sprouts (e.g. aspen after clear cutting), or advance regeneration (e.g. sugar maple or other shade-tolerant species after a tornado).
- The individuals are part of an age group called a cohort.
- This stage lasts from the time of stand-replacing disturbance until the new cohort forms a continuous canopy and trees begin competing with each other for light and canopy space.

Stage 2. Stem exclusion. *During this stage, the canopy is dense enough to prevent new saplings from growing into the canopy – there is no space available for new canopy trees.*

- The canopy continues to have only one dominant cohort, with a uni-modal dbh distribution. The canopy upper surface is relatively smooth.
- Competition among trees is intense and density-dependent self-thinning is the major cause of mortality.
- Crowns are small enough so that when one tree dies, the other trees are able to fill the vacated space in the canopy by expanding their branches horizontally. This situation continues for 100–150 years in northern hardwoods and red or white pine stands, but may last only 20 to 40 years in some aspen and jack pine stands.

Stage 3. Demographic transition. *At this point, a stand undergoes demographic transition from one cohort of trees in the canopy to more than one cohort. There may be a wave of high mortality as many trees reach senescence at the same time.*

- The crowns of the trees are now large enough so that when one dies the surrounding trees cannot fill the gap. As a result, a new cohort of trees has space to enter the canopy. The dbh distribution has the remnants of an old unimodal peak in larger size classes and a new peak in the small size classes. This can be called a 'compound dbh distribution'.

- If the stand was originally composed of a pioneer species such as paper birch, shade-tolerant trees such as sugar maple, hemlock, beech or spruce and fir may begin entering the canopy.

- There are more gaps in the canopy and more light on the forest floor than during the stem-exclusion stage. Mid-tolerant trees such as basswood, green ash, yellow birch and white pine may be able to enter the canopy through some of the larger canopy gaps.

- Mortality undergoes a transition from mostly density-dependent self-thinning to mostly density-independent mechanisms such as senescence and blowdown due to weakened wood caused by heartrot or disease. Dense clusters of young trees in large gaps still undergo self-thinning (i.e. the whole stand development sequence may repeat itself at the neighborhood spatial scale within gaps).

- The stand begins to take on 'old growth' characteristics, with large rotten logs on the forest floor, many sizes of trees, and an uneven canopy surface.

- This stage lasts from the time the first trees younger than the disturbance cohort are able to grow into the canopy until the disturbance cohort no longer has a significant presence in the stand.

Stage 4. Multi-aged. *At this point demographic transition to uneven-aged status is complete, and the forest has many age classes and size classes of trees in the canopy. There may be few or no remnants left from the original cohort. Mortality is continuous at a relatively low level, caused by death of single trees or small groups of trees.*

- The dbh distribution is characterized by many small trees, with a steep decline in number of trees until the middle size classes, where the decline becomes more shallow or levels off, followed by another sharp decline in the largest size classes. The dbh distribution results from a high rate of mortality for small trees (which are undergoing self-thinning in gaps), low mortality for middle-aged trees, and high mortality for large, senescent trees.

- Changes in species composition, development of old growth characteristics, and density-independent mortality with some self-thinning in gaps described for the reinitiation stage continues.

- This stage will last until another stand-replacing disturbance occurs.

Many readers will undoubtedly notice that I have changed the last two stages from Oliver's scheme (Oliver 1981, Oliver and Larson 1996). What I call 'demographic transition' was referred as to as 'understory reinitiation' by Oliver. I made this change because in the Great Lakes Region, this stage has a lot of variability among forest types. Some forests have an understory layer of tree seedlings in all stages of development. However, those seedlings rarely have any chance at all of growing into the mid-story and then capturing a gap as long as the stand is in the stem-exclusion stage. Therefore, the understory is not really reinitiated at a given point in time. What I call the 'multi-aged' stage of development was referred to as 'old growth' by Oliver. The problem with 'old growth' is that it has become a political term that has many different definitions used by managers in different countries and in different states throughout the United States. Environmentalists have tagged any forests with large trees, or old trees (especially older than people's maximum lifespan), as 'old growth'. This includes stands of long-lived species such as oak, maple, and white pine in the Great Lakes Region that are 120–150 years old but still in the stem-exclusion or demographic-transition stages of development. Therefore, to avoid confusion with these many political definitions of old growth, I have introduced the term 'multi-aged' forest. This term is also a good descriptor of the forests in question, since we are always talking about forests that have undergone transition from even-aged to uneven-aged. Multi-aged says exactly what it means with no ambiguity.

Stand development can also be placed on a continuous scale. For example, in sugar maple forests of Upper Michigan, modal stand diameter, proportion of total crown area in large trees (>46 cm dbh), and ratio of mature (26–45.9 cm dbh) to large trees are useful measures for quantifying stage of stand development (Lorimer and Frelich 1998). Figure 4.2 shows the dbh distribution for six stands ranging from young even-aged to nearly balanced all-aged. Stands with $<45\%$ of the crown area in large trees have unimodal diameter distributions, and as the ratio of large to mature trees exceeds 1.5, the form of the diameter distribution changes from unimodal to multi-modal, irregular, or descending monotonic, and the ratio of large to mature trees also increases throughout this sequence (Figure 4.2). Thus, these scaling factors allow the placement of stands on a developmental continuum that continues further than a chronosequence of stand ages. This can be an advantage because stand age becomes difficult or impossible to determine at or beyond the demographic transition stage of development.

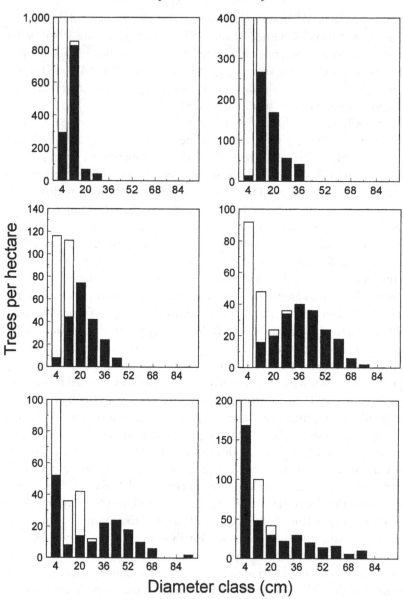

Figure 4.2. Diameter distributions of sugar maple from six stands in the Porcupine Mountains and Sylvania Wilderness Areas in Upper Michigan, arranged in order of stand structural chronosequence (top left, right, middle left, right, bottom left, right). Canopy trees and gap saplings are shown as shaded bars and suppressed trees are represented by open bars. After Lorimer and Frelich (1998).

Succession, fluctuation, and stand development

Another aspect of stand development to keep in mind is its relationship to changes in species composition, commonly referred to as succession or fluctuation. I use the common definition of succession: a directional change in species composition over time, where one species or group of species replaces another. Changes in composition of lesser magnitude than replacement are termed 'fluctuation', where the relative proportion of two or more species shifts over time. Fluctuations may be major (e.g. the ratio of maple to hemlock shifts from 25:75 to 75:25) or minor (e.g. the ratio of maple to hemlock shifts from 40:60 to 60:40).

What about the relationship between changes in composition and stand development: do they necessarily parallel one another? As shown in the five case studies below, sometimes stand development is accompanied by succession, but often it is not. There is often the incorrect assumption by many that stand-replacing disturbance always initiates a new successional sequence. Severe disturbance always initiates a new stand development sequence, which has sometimes in the past been called 'physiognomic succession' to differentiate it from the classical notion of succession that refers strictly to changes in species composition over time.

Before we launch into case studies that spell out the relationships between species composition change and stand development, it is necessary to have an understanding of what is meant by 'a directional change in composition over time', as was stated in the definition of succession above. The four stages of stand development described above are definitely directional, but what about successional development? Predicting the direction of succession in forests has been a major focus of ecological research as long as the science of ecology has existed (e.g. Gleason 1927, Clements 1936, Watt 1947, Curtis 1959, Drury and Nisbet 1973, Horn 1974, West et al. 1981, Peet 1992, Frelich and Reich 1995b). I have detected five major models of successional direction in the literature (Figure 4.3).

Cyclic model

Succession starts with composition state A, then proceeds to state B, C, etc., eventually returning to state A. Cyclic succession was first proposed by Watt (1947) and has been one of the major models employed by

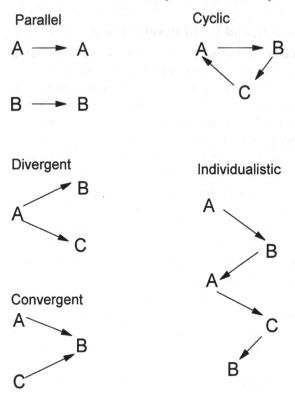

Figure 4.3. Five models of successional direction.

ecologists. An example is aspen invading an area after severe fire, followed by red maple and other species of intermediate shade tolerance, then by shade-tolerant northern hardwoods and hemlock, which then persist until another fire resets the cycle to aspen (Lorimer 1977, Whitney 1986, Frelich 1992).

Convergent model

In the classic model of Clements (1936), vegetation in two or more states (A and B) converges to state C over time. Two adjacent post-fire stands dominated by early or mid-successional species such as aspen and white pine, respectively, which both succeed to shade-tolerant sugar maple and hemlock, provide an example.

Divergent model

One community (state A) diverges into two or more states (B, C, etc.), over time. The divergence involves feedback switches that magnify initial minor differences, and, once the differences are large, allow their perpetuation (Wilson and Agnew 1992). An example of this process could be a post-fire aspen stand that could easily succeed to pine, oak and maple on three adjacent sites. These differences in trajectory could be caused by soil differences that favor different species or by differences in seed availability.

Parallel model

Communities in states A and B each undergo stand-replacing disturbance, and each returns to the same state shortly after the disturbance. This is what some would call no succession, or is used to describe return to original condition after a short period of transient dynamics. Parallel succession can occur in the North American boreal forest, where species such as jack pine or black spruce can dominate adjacent stands. When stand-replacing fires occur in such areas the species in the pre-fire stand are the only seed source, and the post-fire stand maintains the same tree composition (Dix and Swan 1971, Heinselman 1981a,b, Johnson 1992).

Individualistic model

Also called multiple pathways of succession (Cattelino *et al.* 1979), individualistic succession occurs when stochastic variables, such as timing of major seed crops of the important tree species, droughts, disturbances, and other factors interact to produce multiple pathways of succession at different times at the same location. This model emphasizes continuous change and there may not be a stable endpoint. For example, a gap in the forest canopy formed in one decade could be filled by paper birch merely because paper birch had a good seed year and other species did not. If the paper birch later die of old age during a drought they may be replaced by a species with drought-tolerant seedlings, such as white cedar, if that species happens to be nearby.

Predicting successional direction

There are several difficulties with trying to predict successional direction in a real-world forest. The first of these is obvious in theory but often invisible in the field. Namely, disturbance regimes are often complex, with several different types of disturbances. The result is what I call a 'successional system', with many developmental and successional stages related to each other in a web that may include segments of any or all of the five directional models of succession. An investigator who does not manage to isolate individual segments of the whole system for study will likely end in confusion, since their data will show elements of more than one directional model, which may wash each other out. In my opinion this is why there are so many studies done where large numbers of field plots and variables such as soil type, disturbance type, and stand age, are put into multi-variate 'black boxes', such as multiple regression, ANOVA, factor analysis, ordination, or canonical correlation, and end up with significant results that only explain 5% of the variation. It is nice to know that there is a significant trend in the data, but one also needs to explain differences among plots to have true predictive ability.

Other factors that may obscure our ability to predict direction of succession in forests include lack of spatial context, lack of consideration of spatial scale, and inadequate knowledge of successional mechanisms. Reviews of succession (Drury and Nisbet 1973, Facelli and Pickett 1990) suggest that the spatial context in which succession occurs is not adequately taken into account in many studies. Neighborhood effects, such as seed rain, shading, and nutrient feedbacks to the soil through litterfall, all play a role in determining how succession proceeds underneath the canopy of every tree. The successional direction for an entire stand (1–10 ha scale) is the sum of all these neighborhood trends, and a number of recent studies attempt to integrate these spatial effects into successional studies (e.g. Lippe et al. 1985, Hubbell and Foster 1986, Smith and Huston 1989, Frelich et al. 1993, Frelich and Reich 1995b). Research on forest succession often concentrates on processes that occur at a single spatial scale. For example, a large number of papers have been published that examine individual tree gaps (e.g. Runkle 1982) or that examine stands (e.g. Grigal and Ohmann 1975) or that analyze landscapes (e.g. Payette et al. 1989, Dansereau and Bergeron 1993). An example of the data-interpretation difficulties this may create occurs for fir–spruce–birch forest reported by Buell and Niering (1957). This forest may be a uniformly mixed forest, with individual trees of different species next to each other (the result of convergent succession), or a series of small mono-dominant stands (the result of divergent succession). However, since no spatial data are presented, this is impossible to ascertain.

Case studies of forest development and succession

Case study 1: The birch–white pine, hemlock–hardwood successional system

The persistence and widespread success of white pine in presettlement forests of eastern North America, and its dependence on fire, is a paradox because white pine does not have adaptations to fire possessed by the classic fire-adapted species. Although white pine responds favorably to exposed mineral soil seedbeds and high sunlight after fires, it does not have serotinous cones like jack pine and black spruce, the ability to sprout vegetatively after fire like aspen and birch, or widespread and abundant seeds nearly every year like aspen and birch. White pine does possess the compensating life-history characteristics of long lifespan, ability of mature individuals to survive surface fire, moderate tolerance to shade, and the ability to grow in poor environments, such as riverbanks and rock outcrops, giving it a permanent refuge (at least it was permanent before humans removed these refuges) from disturbance and competition by shade-tolerant species (Figure 4.4).

Stand development after severe fire
There is a delicate balance between the abundance of white pine and fire: too much fire and the system shifts to mostly paper birch; too little fire and white pine succeeds to sugar maple–hemlock forest. Thus white pine has the curse of mid-successional species in that it can't get along with fire and can't get along without it. Severe natural fires ignited by lightning occur in the range of white pine during spring, or late summer/autumn, after 2–3 months of drought at the sub-continental scale (Haines and Sando 1969, Heinselman 1973, Cwynar 1977). White pine is most abundant in areas where the rotation period, or mean interval between severe fires, is 150 to 300 years (Frelich 1992). If the fire cycle becomes more than 300 years for any reason, then hardwoods invade and change the fuel type in such a way that fires become much less common. Windthrow and other treefall gaps then become the dominant disturbance type. Forests within the range of white pine experience treefall mortality of 10% or more every 70 years on average, and they experience stand-leveling windthrow at intervals of 1000 to 2000 years (Lorimer 1977, Canham and Loucks 1984, Whitney 1986, Frelich and Lorimer 1991a). Windthrow is generally not favorable for white pine establishment, although a few white pine generally occur in post-blowdown stands. Only one instance of heavy recruitment of white pine after windthrow is mentioned by the literature (Hough and

Figure 4.4. Refuges from disturbance for white pine: riverside (foreground), and rocky ridges (background, upper left). These refuges are nearly permanent unless humans cut them down. Photo: University of Minnesota Agricultural Experiment Station, Dave Hansen.

Table 4.1. *Period of peak seedling recruitment for white pine (starting the year after fire, and ending at the number of years indicated)*

Peak recruitment (yr)	Associated species[a]	Location	Reference
10	Red maple, gray birch, white ash, cherry	Pennsylvania, USA	Hough and Forbes 1943
20	Hemlock, paper birch	New Hampshire, USA	Henry and Swan 1974
20	Pin oak, bur oak, jack pine	Wisconsin, USA	Frelich 1992
25	Aspen, paper birch, red maple	Minnesota, USA	Frelich 1992
25–30	Aspen, red pine	Ontario, Canada	Cwynar 1977
25–35	Paper birch, red maple, hemlock	New Hampshire, USA	Foster 1988a
20–40	Chestnut, red maple, red and white oak	Pennsylvania, USA	Hough and Forbes 1943
40	Paper birch, aspen	Quebec, Canada	Maisurrow 1935

Note:
[a] Species recruited at the same time.

Forbes 1943), although many instances of heavy recruitment after fire are documented.

White pine has abundant seed crops every 3–5 years and very slow growth during the first 5–10 years after germination. In addition, after severe crown fire only a few surviving white pine remain in most forest stands. Therefore, seedling establishment by white pine in post-fire stands usually takes place over 20–40 years under a faster growing or earlier established canopy of aspen, birch, red maple or oak (Table 4.1). Some white pine seedlings may be present right away in the stand initiation phase of development if restocking by other species is slow (Figure 4.5). The shade cast by aspen or oaks is not as dense as that in hemlock–hardwood stands, so that growth of the mid-tolerant white pine seedlings is good – usually 0.3–0.6 m per year. Near the end of the stem exclusion phase of development (which varies from 40 to 80 years of age), many of the white pine are tall enough to take positions in the upper canopy as the birch begin to die (Figure 4.5).

During the demographic-transition stage white pines enter the canopy

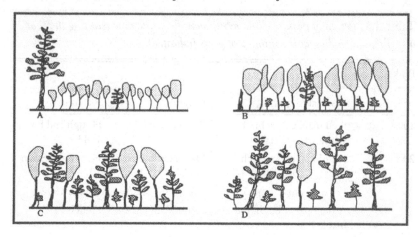

Figure 4.5. Succession and stand development after stand-killing fire in a paper birch–white pine successional system. Initiation (A), Stem exclusion (B), Demographic transition (C), and Multi-aged stages (D). From Frelich (1992).

at different times, as gaps caused by the death of mature birch appear. In addition, other white pine may have been in the canopy from the early stand initiation phase, and others are suppressed under larger white pines. These processes lead to differential growth rates and the development of a hierarchy with rapidly growing canopy-emergent white pine and slower growing trees in intermediate and overtopped crown classes. The diameter distribution becomes bimodal (Figure 4.5).

Two types of low-severity disturbance facilitate the transition from demographic transition to old multi-aged phases of development, which generally occurs between 100 and 200 years of age. White pine is very susceptible to high winds and the larger the tree, the more susceptible it is (Stoekeler and Arbogast 1955, Foster 1988b). Therefore, trees in the canopy emergent position are often toppled by wind and other white pines from a lower position are released into the canopy. A second disturbance may be surface fires at intervals of 20–40 years (Frissell 1973). If shade-tolerant hardwoods are invading the understory they may be killed by surface fires. Also, a few of the large white pine may be killed by surface fires, opening gaps for recruitment of new white pine seedlings and saplings. The gaps caused by wind and surface fire cause a stand to become increasingly multi-aged, with a multi-modal diameter distribution. White pine stands may be maintained in the old multi-aged stage for one to several centuries until another severe crown fire occurs (Heinselman 1981b).

A pattern of cyclic succession

A repeating sequence of severe crown fires with intervening low to moderate-severity surface fires can lead to long-term dominance by white pine in one stand for several thousand years (Davis *et al*. 1994, 1998, Figure 4.6). There would be pulses of species such as paper birch after each major fire, so that some succession would occur from time to time. When old multi-aged white pine stands are maintained by surface fires, periods of several centuries with little succession are also possible. Other cycles may also occur in which the birch/white pine forest alternates with

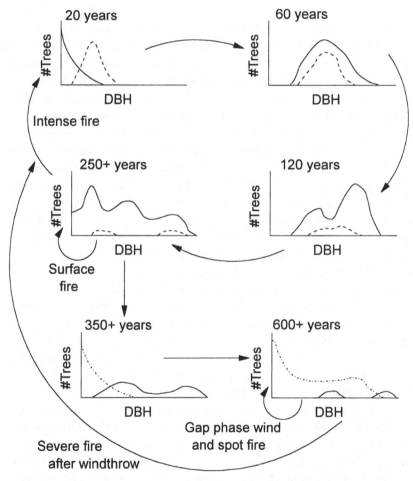

Figure 4.6. Cyclic succession in the birch, white-pine, sugar maple successional system. Diameter distributions at various ages after stand-killing fire are indicated by solid lines, long dashed lines and dash-dotted lines for white pine, paper birch, and sugar maple, respectively. After Frelich (1992).

Table 4.2. *Chance of an individual stand surviving one or more rotation periods under a constant probability of fire with stand age*

Number of rotation periods	Stands surviving
1	36.8
2	13.5
3	5.0
4	1.8

hemlock–hardwood forest (Figure 4.6). A necessary condition for succession from white pine to hemlock–hardwood is several centuries without severe crown fire and infrequent surface fire. Two major factors can cause or accelerate such successional events. First, fire may miss a stand by chance. Under natural conditions disturbances that occur at random with respect to stand age may skip individual stands for one or more consecutive rotations (Van Wagner 1978). For example, if the natural rotation period for crown fire in white pine forest is 200 years, then 14% of all stands may survive two rotation periods (400 years) and a few stands may even survive three or four rotation periods (Table 4.2). This would be enough time for succession to northern hardwoods. Successive windstorms would selectively remove the much taller canopy-emergent pines and release understory shade-tolerant species such as hemlock and sugar maple – a process known as disturbance-mediated accelerated succession (Abrams and Nowacki 1992). The second factor promoting the switch to hemlock–hardwood forest is species composition–fuel feedbacks. If a white pine stand is missed by fire for 2–4 centuries, and species like hemlock and sugar maple become a major component of the forest, then the fuels become more difficult to burn, and the rotation period for fires is lengthened. This is consistent with the relative lack of recent burns and abundance of recent windfalls recorded by the nineteenth century land surveyors in northern Wisconsin and Upper Michigan (Bourdo 1956, Canham and Loucks 1984), and with observations that northern hardwood forests are fairly resistant to burning (Stearns 1949, Curtis 1959, Miller 1978, Bormann and Likens 1979). Other factors can also contribute to the paper birch/white pine to hemlock–hardwood successional trend: (1) a coincident change to cooler and wetter climate (Davis *et al.* 1992, 1994, Clark *et al.* 1996); (2) fine-textured soils that favor rapid growth of hemlock and hardwoods; and (3) the presence of topographical firebreaks.

Succession from hemlock–hardwood forest back to paper birch/white pine requires a severe fire (Figure 4.6). However, hemlock–hardwood stands have moist foliage and relatively low canopy bulk density, and they cannot carry a crown fire. Therefore this successional event will be rare and be facilitated by windfall–fire interactions. Although severe fires are not common in hemlock–hardwood forests (Frelich and Lorimer 1991a), slash from a stand-leveling windfall event can be quite extensive and provide the necessary fuel for severe fires (Stearns 1949, Whitney 1986, Foster 1988a, Lorimer 1977). The post-fire successional sequence commonly included paper birch and white pine. Drought is a necessary factor for windfalls to burn. At any one time on a natural landscape there are areas of recent windfall. However, they are much more likely to burn if a drought occurs. A coincident change to a warmer/drier climate, sandy soil, and lack of topographical firebreaks will aid this successional sequence and allow the return of the birch/white pine forest.

Succession, stand development and wind in the hemlock–hardwood forest
These forests have a tendency to undergo divergent succession, segregating into hemlock-dominated and sugar maple-dominated stands (Davis *et al.* 1994; see Chapter 6, 'Sylvania case study', for more detail). Once these hemlock or sugar maple-dominated stands are established, however, many centuries can go by without any succession within stands (Davis *et al.* 1998). The previously described double whammy of windfall followed by fire, causing the birch–white pine–hemlock–hardwood cycle, does not occur very often. Instead, most of the dynamic action is in constantly changing stand structure (Figure 4.7).

These forests experience many single treefall gaps, a moderately large number of disturbances that kill 5–40% of the trees, and very few stand-leveling wind events. The medium disturbances have a greater cumulative impact on the forest than either single tree gaps or stand-leveling wind (see Chapter 6, 'Wind regimes and landscape structure'). As a result, there are few even-aged stands, and few all-aged stands, but many very complex multi-aged stands. Therefore the basic stand development scheme needs some refinement to take into account many different types of multi-aged stands. There does not seem to be a straight-line sequence of stand development, but rather a 'web of development', with many interlinked types of stands and developmental pathways. Frelich and Lorimer (1991b) identified eight different structural types, all of which work equally well for hemlock and sugar maple stands, and can be viewed as variations on the central theme of the four basic stand development stages (Box 4.1, Figure 4.8).

Figure 4.7. Partial blowdown and recovery in an old multi-aged hemlock–hardwood forest in the Porcupine Mountains, Upper Michigan. Photo: Lee Frelich.

BOX 4.1

Stand development stages for sugar maple–hemlock forest in Upper Michigan. Each stage entry is followed by the equivalent stage from the general stand development scheme (Figure 4.1), a general description of how the stage is formed and what type of structural evidence might place a stand in a given stage. The detailed classification rules are given in Frelich and Lorimer (1991b), and were formulated so that all stands on the landscape could be assigned to one of the eight stages. Remember that disturbance events can cause a discrete change in developmental category, but that growth after a disturbance is a continuous process. Therefore, the classification rules provide somewhat arbitrary boundaries between stages as stands develop.

Sapling stand (equivalent to stand initiation):
  Very young relatively even-aged stands created by intense blowdown that has removed the former canopy, all at once, or in two or more episodes within two decades. A substantial majority of canopy area is occupied by saplings (trees 0–10.9 cm dbh) and poles (trees 11.0–25.9 cm dbh), with aggregate crown area of saplings greater than that of poles.

Pole stand (early stem exclusion):
   Young even-aged stand that develops from growth in a sapling stand. A substantial majority of the canopy area is occupied by saplings, poles, and mature trees (trees 26.0–45.9 cm dbh), with aggregate crown area of poles greater than that of saplings or mature trees.

Mature stand (late stem exclusion):
   Middle-aged, even-aged stands resulting from growth of a pole stand. A substantial majority of the canopy area is occupied by poles, mature trees, and large trees (>46.0 cm dbh), with aggregate crown area of mature trees greater than that of poles or large trees.

Multi-aged break-up stand (demographic transition):
   Typically an old, first-generation stand after canopy-leveling windstorm that has more mature and large trees than a mature forest, but does not yet have a broadly uneven-aged structure. These stands result from growth and development in mature stands.

Multi-aged pole stand (a subdivision of the multi-aged stage):
   A stand with size structure similar to that of a pole stand, but not originating from a sapling stand. Unlike a pole stand, widely scattered age classes are present. Multi-aged pole stands result when a moderately severe windstorm reduces the stature of any of the other multi-aged stand types or a steady-state stand, by removing most of the large and mature trees and leaving behind mostly pole-sized trees. A windstorm that creates multi-aged pole stands does not quite reduce stand stature enough to place the stand in the sapling stage.

Multi-aged mature stand (a subdivision of the multi-aged stage):
   A stand with structure similar to that of a mature stand, but not originating from a pole stand. Unlike a mature stand, widely scattered age classes are present. Multi-aged mature stands result when a moderately severe windstorm reduces the stature of old multi-aged stands or steady-state stands, by removing large trees while leaving mature trees intact. This stage can also result from continued growth in a multi-aged pole stand.

Old multi-aged stand (a subdivision of the multi-aged stand):
   A stand that has gone more than 250 years since the last canopy-leveling disturbance and has a canopy dominated by mature and large trees. Usually at least 50% of the canopy area is occupied by large trees. Generally, several to many widely scattered age classes are present that are unequal in area occupied. Old multi-aged stands result from growth of multi-aged break-up, multi-aged pole, or multi-aged mature stands, or from partial canopy removal in a steady-state stand.

Steady-state stand (a subdivision of multi-aged stand, also sometimes called balanced all-aged stand):
   A stand that has gone more than 250 years since the last major disturbance and has experienced continuous formation of small scattered treefall gaps over the last 250 years, so that the area of new gap formation has been relatively constant over time.

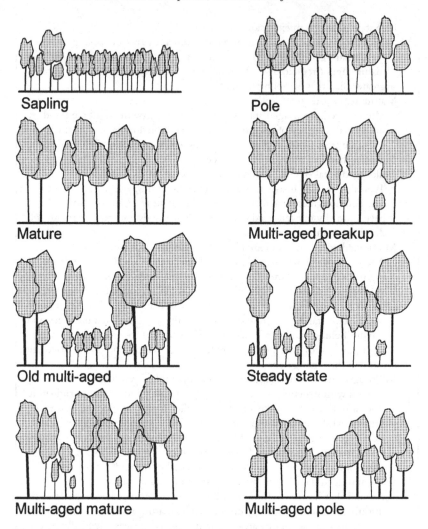

Figure 4.8. Eight stages of stand development in hemlock–hardwood forest. See Box 4.1 for definitions.

Case study 2: The southern-boreal jack pine–aspen, spruce–fir–birch–cedar successional system

Basic development sequence

These jack pine, aspen and black spruce dominated boreal forests are as similar to those of central Canada as one can find in the lower 48 states. Within the primary forest remnants of the Boundary Waters Canoe Area Wilderness (BWCAW), the natural processes of disturbance and forest

Figure 4.9.   Stand initiation in an upland jack pine–black spruce stand after a high-severity crown fire in Minnesota's Boundary Waters Canoe Area Wilderness. Note that much of the soil was a moss mat that burned away, exposing bare rock. Photo: University of Minnesota Agricultural Experiment Station, Dave Hansen.

development still occur. These forests on nutrient-poor shallow soil over granitic bedrock originate after severe crown fires (Figure 4.9). Under the pre-European natural disturbance regime, severe crown fires had a rotation period of approximately 50–70 years (Van Wagner 1978, Heinselman 1981b).

Sometimes fires occur in young stands (<10 years old), an age at which the conifers do not produce much seed, and aspen comes to dominate the post-fire stand (Heinselman 1973). Otherwise, if fires occur when jack pine and black spruce are old enough to produce abundant seed, then those species dominate the post-fire stand. Jack pine has serotinous cones that remain closed until scorched, and after a crown fire, seeds rain down by the millions on every hectare. Black spruce has semi-serotinous cones; a few open every year, but the majority are stored in the canopy and open after fire. Surveys of seedling densities after fire indicate that densities of 100 000 to 300 000 seedlings per hectare are common 2–3 years after fires (Heinselman 1981b). This initiation phase of development lasts less than a decade, by which time a very thick young forest closes the canopy. After that, stem exclusion begins and continues until

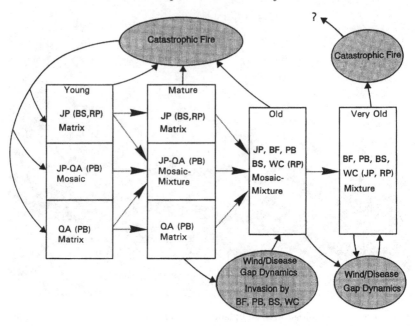

Figure 4.10. Successional and developmental web for the near-boreal forest of the Seagull Lake area in northern Minnesota's BWCAW. Dotted arrows show pathways caused by growth and self-thinning. Solid arrows show pathways caused by exogenous forces shown in the shaded ovals. BS, black spruce; PB, paper birch; WC, white cedar; BF, balsam fir; QA, quaking aspen; RP, red pine; and JP, jack pine. From Frelich and Reich (1995a).

approximately age 100 (Frelich and Reich 1995b). Demographic transition occurs between age 100 and 150 as large gaps form when small groups of jack pine die. Balsam fir, white cedar and paper birch invade these gaps, and black spruce also increases its importance within stands during those years. Finally, when the original jack pine have nearly all died, after age 150, stands become a mixture of the 'climax species group' of black spruce, balsam fir, paper birch and white cedar, a more complex group than the hypothetical spruce–fir often mentioned in the literature (Frelich and Reich 1995b, Figure 4.10).

All directions of succession occur in this forest

Can parallel, convergent, divergent, cyclic and individualistic succession all occur at the same time in one forest? Yes, but it all depends on how you look at the system. Let's examine the system at several spatial scales and also compare stands of many different ages at one point in time and

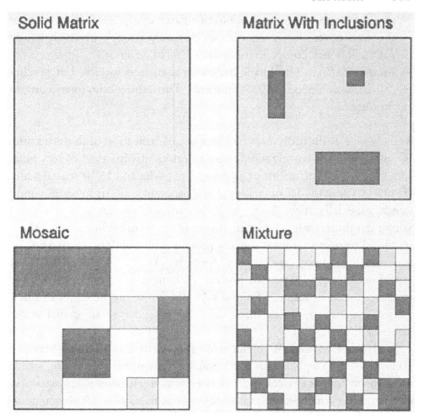

Figure 4.11. Idealized spatial structures in the near-boreal forest. From Frelich and Reich (1995a).

the same stands at several points in time. Frelich and Reich (1995a) accomplished this suite of analyses by mapping all trees within some stands, reconstructing stand history since the last fire, and examining the spatial structure and composition of stands of widely varying ages (origins from 1801 to 1976) on aerial photographs of the same forest taken in 1934, 1961 and 1991. Air photo interpretation proceeded by classifying forest blocks of 1 ha, 4 ha, or 16 ha in size on all three air photos into one of the following four spatial structures (Figure 4.11):

1. *Solid Matrix (MAS)*. Block is nearly completely dominated by one species (makes up >90% of tree cover).
2. *Matrix with Inclusions (MAI)*. One mono-specific patch comprises at least 50% of the area.
3. *Mosaic (MOS)*. The block has mono-specific patches (>0.25 ha in

area) of two or more species. No one patch comprises 50% or more of the tree cover; however, one species may occupy more than one patch, and may comprise more than 50% of cover.

4. *Mixture (MIX)*. The block has more than one species, but patches >0.25 ha comprise <10% of the area. Thus, the species form a diffuse mixture.

The above classification leads to a simple quantification of forest texture. Results reveal that young stands are predominantly matrices of jack pine, sometimes with inclusions of aspen at 1 ha, 4 ha and 16 ha spatial scales (Figure 4.12). Mosaics of jack pine and aspen also occur in some young stands, especially at the 16-ha scale. Mixtures of species are rare among young stands. In contrast, mature forests in the stem exclusion stage show that black spruce that were initially present in the understory make their way into the canopy, so that some forest blocks at 1 ha, 4 ha, and 16 ha spatial scales move from the matrix with inclusion to the mosaic or mixture spatial structure. Old and very old forests are mixtures of black spruce, balsam fir, paper birch, and white cedar at all spatial scales (Figures 4.10, 4.12).

Analysis of transitions of the same forest block over time (between 1934 and 1991, or between 1961 and 1991) confirm that young stands move from solid matrices and matrices with inclusions to mosaics and mixtures over time. Stands that started out as mosaics in 1934 change to mixtures by 1991, and finally, all observed transitions in very old stands result in the continuation of the mixture spatial texture. Of those 1-ha blocks of forest that start out as solid matrices in 1934, some stay in the same category until 1991, while others change to matrices with inclusions or to mixtures. These varying pathways depend on the erratic distribution of canopy-gap formation, which may miss some 1-ha blocks for decades, hit others lightly creating a few small openings that become inclusions, or create many openings resulting in a mixture. Thus, all blocks that are solid matrices will eventually make their way to the mixture category, but the timing will vary depending on the fine-scale spatial and temporal distribution of canopy mortality. Also, bigger blocks of forest undergo transition from matrix to mosaic to mixture slightly faster than smaller blocks. This is expected because as canopy openings cause transition of bigger blocks to mosaics and mixtures, there could be 1-ha blocks embedded within a large block that by chance escape canopy disturbance and temporarily remain a solid matrix (Figure 4.12).

Spatial patterns in an example of very old forest where all trees were

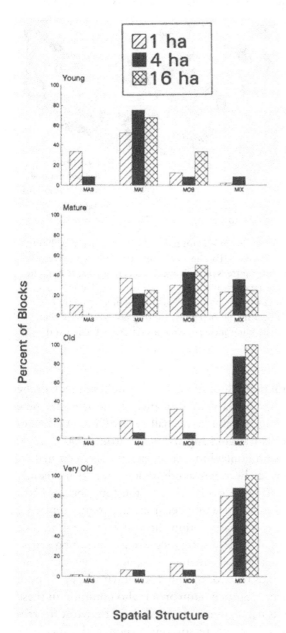

Figure 4.12. Spatial structure by stand age and spatial scale in the near-boreal forest. Percentage of blocks (1, 4, and 16 hectares in size) that are solid matrices (MAS), matrices with inclusions (MAI), mosaics (MOS), or mixtures (MIX) are shown for young (0–40 year old), mature (41–100 year old), old (101–150 year old), and very old (>150 years). From Frelich and Reich (1995a).

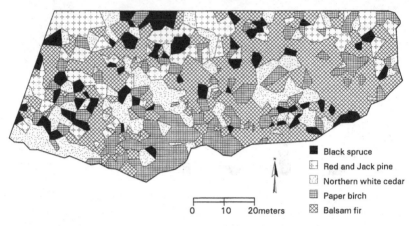

Figure 4.13. Patch map of a very old fire-origin red and jack pine stand well into demographic transition at age 192 years. Patches with different shadings represent the area occupied by canopy and gap trees of each species receiving direct sunlight from above. The patches were formed from a stem map by making Tiesen polygons around each canopy tree, using the perpendicular bisectors of line segments from the subject tree to each of the near neighbors. The resulting polygons were then merged wherever two trees of the same species shared a side. After Frelich and Reich (1995a).

mapped are complex, with multiple and single-tree patches of each of the four climax species, along with a few remnants of the original pine cohort (Figure 4.13). Average patch size is small (35 m²), but 20–40% of area is in the largest patches (ranging approximately from 1000 to 2300 m²). Surprisingly, there is no contagion within species between mature trees and seedlings in gaps, or between standing dead trees that recently died and saplings in gaps (Figure 4.14). In fact, the four species black spruce, balsam fir, paper birch and white cedar are apparently replacing each other on a patch-by-patch basis within the stands (Table 4.3). The patches are renewed by small-scale disturbance events, including senescence of old birch, summer thunderstorm winds, and heavy wet snow accompanied by strong winds that commonly take down small groups of white cedar and black spruce. Spruce budworm is also common in these older stands, and despite its name, infests mostly balsam fir. Most fir trees do not live beyond 40 years for this reason, even though the species is capable of living 200 years. There are virtually always young fir in the forest understory that are not killed by budworm, and when they mature they are infested themselves (Heinselman 1981a).

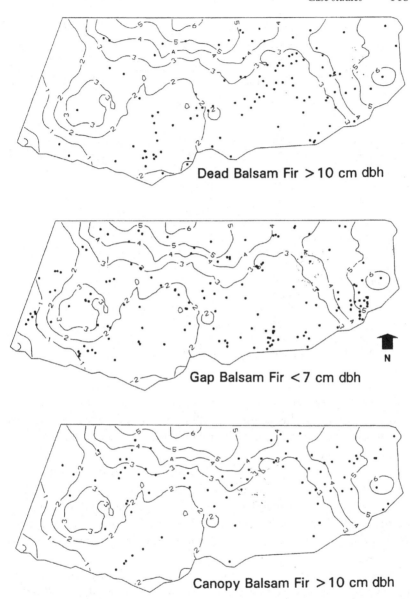

Dead Balsam Fir > 10 cm dbh

Gap Balsam Fir < 7 cm dbh

N

Canopy Balsam Fir > 10 cm dbh

Figure 4.14. Lack of contagion between adults, recently dead trees and saplings in gaps for balsam fir. Contours show elevation in meters, starting from the lowest point on the plot, which is 126 meters long east-to-west. After Frelich and Reich (1995a).

Table 4.3. *Transition probabilities of from species-to-species on two plots*

Transition from (n):	Transition to:			
	Black spruce	Balsam fir	Paper birch	White cedar
	Transition frequency			
Threemile Island (192 year old stand)				
Jack/red pine (77)	0.286	0.195	0.325	0.195
Black spruce (43)	0.326	0.209	0.349	0.116
Balsam fir (49)	0.286	0.327	0.224	0.163
Paper birch (11)	0.181	0.545	0.273	0.0
White cedar (12)	0.083	0.0	0.167	0.750
All (192)	0.276	0.240	0.292	0.193
Fishook Island (127 year old stand)				
Jack pine (20)	0.600	0.050	0.300	0.050
Black spruce (44)	0.610	0.023	0.364	0.0
Paper birch (1)	0.0	0.0	1.0[a]	0.0
All (65)	0.600	0.031	0.354	0.015

Note:
[a] Note sample of 1.
Source: After Frelich and Reich (1995b).

Now, in light of the evidence just presented, we return to the question at the beginning of this section: can all directions of succession occur in one forest at the same time? The following paragraphs summarize the empirical evidence.

Parallel succession clearly occurs when stands that still have substantial numbers of jack pine or aspen burn (Heinselman 1973, 1981a,b, Frelich and Reich 1995b, and Figure 4.10). Based on the landscape-level analysis of the BWCAW by Heinselman (1973), parallel succession in tree-species composition occurred at all spatial scales from individual tree to landscape.

Divergent succession is occurring at fine-grained spatial scales, such as the individual tree scale and the 0.25 ha scale (Figure 4.13). Whether reconstructing history of one stand via increment core analysis or following stands over time on sequential air photos, all young stands eventually progress to mosaics or mixtures that can be viewed as very small stands of individual species.

Convergent succession is occurring in the very same stands experiencing divergent succession, at the exact same time, at spatial scales of 1 hectare or more (Figure 4.12). The mixture or mosaic of black spruce, balsam fir, paper birch, and white cedar looks pretty much the same on one hectare as on another.

Cyclic succession occurs when old stands burn, because a substantial proportion of late-successional species invades the jack pine or aspen, before burning and returning to those early successional species (Figure 4.10). The fact that some species such as balsam fir and white cedar do not invade many stands for 100 years or more after fire, and are removed at the time of fire, indicates that cyclic succession occurs.

Succession is individualistic in these forests, because the timing of transition from a matrix of jack pine to a mixture of spruce, fir, birch and cedar depends on the timing of death of the jack pine, which in turn depends on the random nature of windthrow that creates openings in the canopy. The timing of invasion by late-successional species also depends on the distance from surviving seed sources after severe fires, which varies tremendously throughout the BWCAW. A third factor leading to individualistic succession is the random timing of severe crown fires. If a stand burns when young, parallel succession is sure to occur. If it burns when old, cyclic succession occurs. If it burns when very old, and there is no jack pine left, a different as yet unobserved trajectory will occur.

Near-boreal forest summary

A scientist studying these forests with some sort of philosophical agenda as to which models of succession are most elegant or theoretically attractive could get the data to fit any successional model they wanted by adjusting the temporal and spatial scale of analysis. Conversely, the confused scientist who doesn't understand the scale-dependent nature of this forest is likely to remain confused after analyzing the data. I recommend multi-scale, multi-temporal analyses of succession because it appears that the true picture of succession can only be arrived at in that way.

One implication of the study of stand response to disturbance in the near-boreal forest is that stable oscillation between successional states is unlikely. Ludwig *et al.* (1997) describe a boreal forest where preferential browsing of aspen and birch in young stands and poor reproduction of aspen under its own canopy lead to succession to fir, which in turn increases the density of fir foliage in the canopy until budworm has an ideal breeding environment, and an outbreak occurs which kills much of the canopy. Then fire occurs and the system goes back to the aspen–birch stage. However, when disturbance severity and timing is the main driver of forest change – such as in the near-boreal case study –

such stable-limit cycles are unlikely because there is no reason that disturbances of similar severity should return on a regular basis. Forest-fire intensity, or the rate at which heat energy is released per unit time, is very dependent on degree of dryness of fuels and the weather at the time of the fire in boreal forests (Johnson 1992). Fire intensity has a large influence on whether a fire will stay on the ground or enter the crowns within a closed-canopy conifer forest, and this in turn determines whether the canopy trees are killed. Thus, fire severity is extremely variable from one event to the next – literally as variable as day-to-day weather. Windstorms have a similar variability in degree of blowdown caused, depending on wind speed and condition of stands hit by the wind (Frelich and Lorimer 1991a). Budworm damage to balsam fir in the near-boreal forest depends on stand composition at the stand and landscape levels (Bergeron *et al.* 1995), thus leading to a wide range of tree mortality among stands. In addition, with an equal chance of burning across stands of all ages, the timing of natural disturbances in the near-boreal forest is very erratic for a given stand. The chance of stand-killing fire does not increase with age, so that the cycle does not necessarily have to go back to the early-successional stage within a certain time frame. Paleoecological analyses of Swain (1973) and Bergeron *et al.* (1998) confirm that climate change, by regulating the disturbance regime, has governed the balance between the early-successional species group jack pine–aspen, and the late-successional species group white cedar–spruce–fir–birch over the last several millennia in the near-boreal forests of Minnesota and southern Quebec. The spatial structure of the forest at several scales also influences the chances of a regularly oscillating system. If a near-boreal forest is to switch back and forth from jack pine to cedar as the disturbance regime changes, the seed source for species in each forest type must always be present. However, cedar and fir are often removed from stands by severe fires (Frelich and Reich 1995a), and the placement of refuges from fire and size of fire cause large variation in the amount of time it takes for cedar and fir to reinvade after fire, from decades to centuries. Because of this and climate changes, the forest is not likely to cycle through the jack pine and cedar phases on any regular basis.

Case study 3: The near-boreal birch, spruce–fir successional system

This variant of near-boreal forest occurs on sites with better quality soils that are relatively deep and fine textured compared with the jack pine forest (Ohmann and Ream 1971). These are dense forests of mixed aspen, birch, balsam fir, white spruce, and red maple. Good soils favor paper birch and aspen immediately after fire, rather than the conifers. Therefore, these forests were undergoing constant succession in response to fire. Fires remove fir and spruce, replacing them with birch and aspen that dominate the initiation and stem-exclusion developmental phases. During the demographic-transition phase of development (age 50–80 years), balsam fir, white or black spruce and red maple invade from intact forest surrounding the burnt areas, attaining dominance by 80–100 years after fire, and remaining as a multi-aged stand until removed by the next fire. The fir and spruce were often high in density, and able to propagate crown fire as well as jack pine forest. But because they were on less drought-prone soils, they did not burn as often. Average intervals between stand-killing fires were about 100 years (Heinselman 1981a). Successional direction can be both parallel (young aspen stands return quickly to aspen when burned) or cyclic (older stands succeed to fir and then return to aspen when burned).

Case study 4: The near-boreal birch–white and red pine, spruce–fir successional system

Near the borders of lakes, on peninsulas, and on islands within the BWCAW, the severity of fires was often lower than on the contiguous uplands dominated by jack pine, aspen and black spruce. The lake edges are often so rocky that fuel for fires was not contiguous, the humidity was higher, the vegetation in some cases could tap into the water table and remain wet even during droughts, and finally, in some areas, a concentration of large lakes served as fire breaks that interfered with the free movement of crown fires across the landscape. In such areas, crown fires occurred much less often (150–200 years average interval) than in the jack pine forest, and many fires dropped to the ground to become surface fires. Such a disturbance regime, with infrequent crown fires and frequent surface fires, favors the development of red and white pine forests (Heinselman 1973, Frelich and Reich 1995b). Red and white pine can only survive fire as mature adults, meaning that they can only survive surface fire, during which the peak temperature inside the bark does not get high enough to kill living cells, and the foliage is not killed. Saplings are usually killed by fire, since their bark is too thin to provide that much insulation, and the tree can be girdled by fire. Individual mature trees that survive the fire are the source of seed.

If there is a major crown fire, most of the large pines are killed. Paper birch invades rapidly, grows much faster than any pine seedlings present, and dominates for all of the initiation and stem-exclusion developmental phases. The few large pines that survived after crown fire, as well as pines outside the burned area, provide the seed source for a gradual reinvasion of the stand underneath the young birch canopy. This wave of reinvasion by white pine typically takes 1–4 decades (Frelich 1992). As the stand reaches the demographic transition phase of development, white pine break through the birch and begin to dominate the canopy by age 100. At this age, the stands reach a triple-point for potential future development and succession. The stand could have another crown fire and go back to the birch-dominated reinitiation phase. The second possibility is that no fire will occur, leading to succession to shade-tolerant species, mainly black spruce, balsam fir and white cedar (Heinselman 1973, 1981b, Frelich and Reich 1995a). Windstorms hasten succession to spruce and fir because the much taller pines are susceptible to blowdown. The third potential pathway for development of the 100-year-old birch forest with pine breaking through the canopy, is that one or more surface fires will occur. Although the mean age at time of stand-killing fire is 150–200 years, some stands go for 400–600 years without stand-killing fires (Heinselman 1973). White and red pine can be maintained for 600 years or more, as long as surface fires occur regularly. Minor surface fires remove shrubs, invading late-successional tree species, and thick duff, allowing for establishment of new cohorts of young pines. If surface fires continue to occur, the stands become multi-aged, usually with 2–4 main age groups (Frissell 1973, Heinselman 1973, Frelich 1992).

In the rocky lands and sandy lands, white and red pine form stands in areas where fire has less than average presence than the surrounding landscape, due to protection from fire by the landscape and topographical setting. These pines cannot perpetuate themselves without surface fire, and they cannot get along with frequent crown fire either. The fire regime and topographic setting has to be just right, which is why red and white pine occupied a relatively small proportion of Minnesota's forest landscape (about 13%) even in presettlement times (Frelich 1995). The timing of crown fires and surface fires determined whether these forests underwent succession from paper birch to white and red pine, and then on to more shade-tolerant species, or whether they remained multi-aged pine stands for several centuries, or whether they reverted to the jack pine or aspen forest type.

Case study 5: Peatland black spruce

Black spruce has a tendency to occupy wetland sites with sphagnum peat, low pH and low nutrient supply (Heinselman 1963, 1970). Wetland areas that are cut off from ground-water flow get most of their water from rainfall and are the most acidic. These are the raised peatlands that hold onto rain water like a sponge, occupying most of the 'Big bog' just to the west of the BWCAW. Other areas with slowly moving ground water have minerals which neutralize the acid produced by sphagnum mosses, leading to dominance by tamarack or cedar.

These lowland conifer forests did support canopy-killing fires, but only half as often as uplands, so that the average interval between fires was about 150–200 years (Heinselman 1981a). Burning usually did not cause much change in tree composition. The main change that occurs in upland areas after fire is invasion of paper birch and aspen, but these two species do not grow well on peatlands. Succession from one conifer to another on peatlands is determined more by long-term changes (over centuries and millennia) in the thickness of peat, changes in drainage patterns, and climate change than by fires (Heinselman 1963).

Because fires were relatively infrequent in the peatlands, many stands reached old ages. Generally, the stem-exclusion phase lasted until age 150–200 years (Heinselman 1963, Groot and Horton 1994). Stands over 200 years old go through demographic transition and eventually become multi-aged. The developmental changes in this case are not accompanied by any successional change.

Summary: general trends of succession and time lines of stand development

Here I present a quick guide to succession (Box 4.2) in the five forest types discussed earlier in the chapter along with idealized time lines of succession as I believe it would proceed under the historic natural disturbance regime (Figure 4.15). These two features can provide the reader with a condensed tabular and visual summary of the chapter.

BOX 4.2. GUIDELINES FOR SUCCESSIONAL CHANGE

THE BIRCH–WHITE PINE, HEMLOCK–HARDWOOD SUCCESSIONAL SYSTEM

- Windfall–severe slash fire combination converts hemlock–hardwood to birch. This type of event has a frequency under natural conditions in the Lake States of once every 1000–2000 years.

- Birch succeeds to white pine over a 100–200 year period. Surface fire maintains white pine and lack of fire allows succession back to hemlock–hardwood.
- Hemlock–hardwood forest can sustain light surface fires and windthrow (if the slash does not burn) with no change in successional status.
- The dynamics of the hemlock–hardwood forest are dominated by a web of structural changes as the forest responds to repeated low to moderate-severity treefall disturbance.

THE JACK PINE–ASPEN, SPRUCE–FIR–BIRCH–CEDAR SUCCESSIONAL SYSTEM

- Severe crown fire at intervals of 20–120 years allows perpetuation of jack pine and aspen.
- Two severe crown fires within a short time remove jack pine and give dominance to aspen.
- Chance lack of fire for more than 200 years allows succession to spruce–fir–birch and cedar.
- Multi-aged spruce–fir–birch–cedar forests can be maintained by windthrow and gap dynamics until the next fire. They may succeed to aspen and black spruce if burned.

THE BIRCH, SPRUCE–FIR SUCCESSIONAL SYSTEM

- Severe fire with mean rotation of 100 years results in removal of spruce and fir and replacement by birch.
- Chance lack of fire for more than 100 years allows succession to fir and spruce.
- Fir and spruce can be maintained by windthrow and budworm.

THE BIRCH–WHITE AND RED PINE, SPRUCE–FIR SUCCESSIONAL SYSTEM

- Severe fire with mean rotation of 150–300 years results in removal of pine and/or spruce and fir, and their replacement by birch.
- Surface fires with mean rotation of 20–40 years can maintain multi-aged pine.
- Chance lack of all types of fire for >200 years (shorter time if wind helps to prematurely remove pines) allows spruce and fir to replace pine.
- Spruce and fir can be maintained by wind and budworm gap dynamics.

PEATLAND BLACK SPRUCE

- Severe fire results in self-replacement by black spruce.
- Windthrow results in self-replacement by black spruce.
- Succession (if it occurs at all) depends on peatland development and change in drainage patterns.

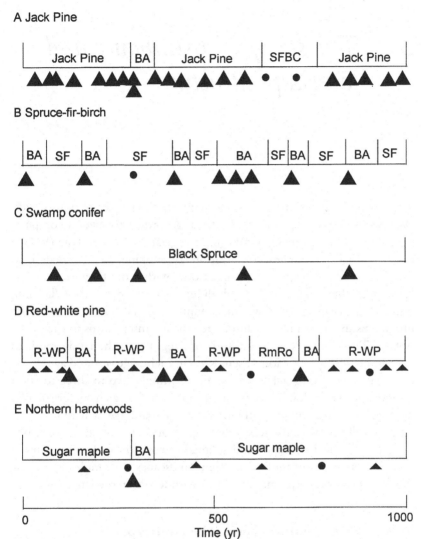

Figure 4.15. Idealized time lines of succession for five forest types under the historic disturbance regime in the Lake States. Large triangles, small triangles, and dots indicate the occurrence of stand-replacing fire, light surface fire, and stand-leveling wind, respectively. BA, paper birch and aspen; SF, spruce–fir; SFBC, spruce–fir–birch–cedar; RmRo, red maple and red oak; R–WP, red and white pine.

5 · The study of disturbance and landscape structure

The forest landscape comprises a collection of many contiguous stands. We may delineate and label these stands according to species composition, developmental stage, stand age, or vegetation growth stage (VGS, a concept which integrates composition, development and age defined below). If the landscape is a complex one, with more than one ecosystem type, there may be several different stand delineation/labeling systems in place, each of which is adjusted to take into account the unique disturbance–physiographic–tree species interactions in each ecosystem. The main effect of a disturbance regime on the landscape is to determine what proportion of stands are in each stage or stand age. If the disturbance regime is stable over sufficient time – two to three rotation periods for the predominant disturbance type – then a characteristic distribution of stands among growth stages or stand age will result. This chapter will start with the relatively simple concept of stand age distributions for simple landscapes and gradually work through to the more complex situations of multi-ecosystem landscapes with multiple successional webs and the presence of old-growth forest on the landscape.

Stand age distribution across the landscape

We can immediately break stand age distributions into two types: stable ones that are capable of perpetuation over time without change in shape, and unstable, or irregular ones. Only flat or monotonically decreasing stand age distributions can be stable (Figure 5.1). Stand age distributions that are unimodal or have two or more large peaks with gaps between are unstable (Figure 5.2). These unstable distributions indicate a changing disturbance regime over time or a landscape that is too small compared with disturbance size to sustain a stable age distribution (see Chapter 8).

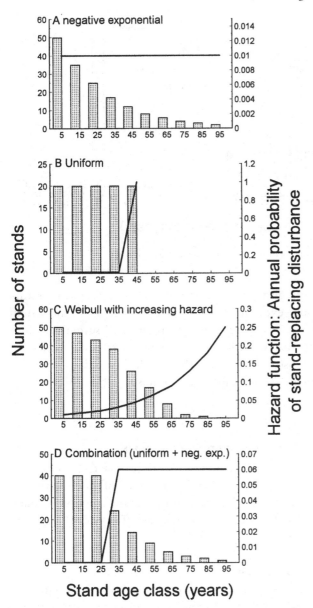

Figure 5.1. Four types of potentially stable stand-age distributions and their hazard functions.

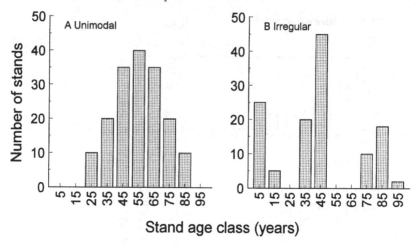

Figure 5.2. Two examples of unstable stand-age distributions.

With an unstable age distribution, one cannot characterize the disturbance regime without extraordinarily detailed knowledge of the system
and very long periods of observation, on the order of several tree lifespans. Small changes (on the order of 50% change in rate of disturbance)
from one stable age distribution to another – called mixed age distributions – may be amenable to analysis (e.g. Johnson 1992, Johnson and
Gutsell 1994).

Stable stand-age distributions and rotation periods

Four types of stable stand-age distributions (or time since disturbance distributions) occur on forested landscapes. Features of interest besides
shape of each distribution, are the shape of the hazard function – which
expresses the chance of disturbance with stand age (Figure 5.1) – and the
fire interval distribution.

The uniform stand-age distribution
This is the dream of many commercial foresters. If a uniform stand-age
distribution can be attained and maintained for a given forest type then a
steady supply of timber may be produced in perpetuity. The statistics are
simple: the rotation period x is the maximum age (or harvest age), and
$1/x$ of all acreage can be harvested annually. The hazard function by stand
age for this distribution is basically at zero until harvest age is reached,
when the chance of disturbance becomes 100% (Figure 5.1B). It is

doubtful that this distribution can occur in any forest ecosystem under a natural disturbance regime.

The negative exponential stand-age distribution
A constant hazard function, or equal probability of disturbance across all stand ages, will result in the negative exponential age distribution (Van Wagner 1978 and Figure 5.1A). Many northern conifer forests have a constant hazard of burning across stand ages, including the white pine and jack pine–black spruce forests discussed in Chapter 4. The formula for the negative exponential is:

$$A(t) = \exp\left(-(t/b)\right) \tag{5.1}$$

where:

t is time in years and b is the rotation period.

The rotation period, b, is also equal to the mean stand age, giving one an easy method of calculating the rotation period from the age distribution (Van Wagner 1978, Johnson and Gutsell 1994). The fire interval distribution is:

$$f(t) = (1/b) \cdot \exp\left(-t/b\right) \tag{5.2}$$

which can be used to predict the proportion of stands that would experience two disturbances within a short time (e.g. how many are hit twice in 10 years?), or what proportion would go for more than one rotation period before being disturbed.

When the cumulative frequency of stand ages is plotted on a semi-log graph, the resulting graph will lie along a straight line if the negative exponential distribution fits. Changes in slope of the semi-log graph – which appears as two or more straight-line segments – delineate changes in the rate of disturbance $(1/b)$. Several of these mixed age distributions from boreal and near-boreal forests are discussed by Johnson (1992).

The negative exponential age distribution has important implications for landscape structure. Since it implies that any given point is hit by disturbance in a random selection process, then by chance some stands will get hit twice in a very short time, leading to the presence of many young stands across the landscape. Sometimes a 'double whammy' effect will occur whereby a second disturbance hits a young stand before the trees are old enough to bear seed. Typically, this leads to presence of deciduous stands within a conifer-dominated landscape (Heinselman 1973). There are many early-successional deciduous species in the genera *Betula* or

Populus around the northern hemisphere that are capable of long-distance seed transport. These species always 'find' disturbed areas where the previous species have been indisposed somehow, and are therefore open for invasion. The double whammy effect on changing composition is minimal in most deciduous forests, where the species can resprout from the stump or root system.

Just as some stands on a landscape with negative exponential distribution of stand ages are hit twice in a short period, other stands will not be disturbed for one or more rotation periods, leading to the presence of old-growth and late-successional species even when rotation periods are short. See Chapter 6 'Fire in the southern-boreal jack-pine forest' for an example.

The Weibull distribution

This distribution is described mathematically as:

$$A(t) = \exp\left(-(t/b)^c\right) \qquad (5.3)$$

where t and b are the same as in the negative exponential (time in years and rotation period), and c is the shape parameter.

The fire interval distribution is:

$$f(t) = ((ct^{c-1}) / b^c) \cdot \exp(-(t/b)^c) \qquad (5.4)$$

When the shape parameter is 1, the Weibull is the same as the negative exponential, with a constant hazard function, and when $c > 1$, the hazard of disturbance increases with age (Figure 5.1C). The Weibull is useful for cases where early-successional stands are not flammable, and flammability increases with age, as when aspen is invaded by conifers (Johnson and Gutsell 1994).

The combination stand age distribution

This is a combination of the uniform and negative exponential. An individual stand goes through a 'waiting period', with very low hazard before it is susceptible to disturbance. After that, there is a transition period, where hazard of disturbance increases, and then a third stage of the distribution follows, where the hazard function stabilizes at a higher level, and stands are disturbed at random in a fashion similar to that described above for the negative exponential (Figure 5.1D). One can estimate the rotation period roughly by breaking the age distribution into two parts – the flat part and the negative exponential part – and calculating the mean stand age only for those stands in the negative exponential part of the distribution. For hemlock–hardwood forest in the Lake States subject to high winds there is virtually no chance of a second stand-leveling blow-down (hazard = 0) for several decades after a first blowdown occurs (Frelich 1986, Frelich and Lorimer 1991a). Once stands are over age 100–130 years, however, stands again reach a constant chance of blow-down, and there is no reason to believe that this hazard function contin-ues to change. Thus, it should generate a negative exponential after age 100–130 years. The combination age distribution has a true change from one fundamental distribution type to another due to inherent properties of the forest and not due to changes in disturbance frequency over time. Therefore, these combination distributions cannot be modeled as a Weibull or a mixed distribution in Johnson's (1992) sense.

Truncated age distributions
Often one cannot obtain stand-age data beyond a threshold age, such as when stands have entered the multi-aged phase of development so that the post-disturbance even-aged cohort is no longer present, or because trees over a certain age are hollow. If there are enough data to determine the shape of the distribution, and it is a negative exponential, then one can look at the slope of the negative exponential, which is directly related to the mean age and rotation period. One can also solve this problem graph-ically by looking for the age threshold at which 63% of all stands are younger, since the 63% point is also directly related to the mean age and rotation for a negative exponential distribution. Of course this strategy will only work when less than 36% of all stands are beyond the truncation point. For extremely truncated age distributions where there is no hint of the true shape of the distribution, other methods for calculating rotation periods are necessary. For truncated stand-age data where the younger stands have a uniform distribution, one cannot simply extrapolate to get the maximum age that represents the rotation period. The shape of the distribution almost certainly changes in older stands, and it will turn out to be a combination age distribution in most cases. The method described below, based on proportion of the landscape disturbed over time, will suffice to estimate rotation periods in cases where age data is truncated.

Variability in rotation periods
It is wise not to take estimates of rotation periods too seriously. I have found it practical to get a consensus from several studies in a given forest type, and then take a two-fold bracket of variability around the consensus value. There is some support for this in the literature. Graphs in Johnson

(1992) show that the rotation period in the BWCAW from 1690 to 1760 was about twice as long as for the period 1760 to 1900. Clark (1988) analyzed charcoal in varved lake sediments in the Itasca area, and found that over the last 750 years, the interval between major fire events varied from 44 to 88 years. Johnson and Larsen (1991) report that fire cycles varied from 50 to 90 years in the Canadian Rockies over the last 400 years, and Bergeron has shown similar variation in boreal forests of southern Quebec (Bergeron *et al.* 1998). When applying rotation periods to estimate the proportion of the landscape in various age classes – whether for purposes of natural area restoration or developing harvesting regimes that mimic natural disturbance – it is important to acknowledge that there is a range of natural variability in landscape structure over time. Unfortunately, one may not know whether they are currently operating under the same natural disturbance rotation periods as those indicated by historical reconstructions.

Estimating rotation periods without stand age distributions

Stand age distributions are not always available. This is especially true for reconstruction of historical disturbance regimes from observations of early settlers and surveyors. It can also be true on landscapes like the Lake States' hemlock–hardwood forest, where stand initiation events have a rotation period so long relative to tree age that any attempt at developing a stand age distribution is severely truncated, to the point where only 10% of all stands may have a detectable stand age (e.g. Frelich and Lorimer 1991a).

   All estimates of rotation periods done in the absence of age distributions are in some way based on the proportion of the landscape disturbed over some period of observation. The basic formula for estimating the rotation period is:

$$\frac{T}{P} \qquad (5.5)$$

where:

T is the period of time during which disturbances are observed (years)

P is the proportion of landscape disturbed

For example, if a landscape were observed for 50 years, and during that time 0.20 of the landscape was disturbed, then the estimated rotation period is $50/0.20 = 250$ years. Note that this calculation assumes that the rotation period is constant over time (i.e. that the time of observation is representative) and that the age distribution of stands is uniform for the period of observation (i.e. there is no depletion of the stands less than 50 years old). If 20% of the landscape was disturbed during a 50-year period on a landscape with a negative exponential age distribution, some of the stands could easily be disturbed twice. Thus, the rotation period would be underestimated by the above formula. The solution for this is to tally multiple disturbances if they occur at one point, or the area of all disturbances, including the area of overlap among disturbances that occur during the time of observation. This may be easy to do from sources such as fire records where each occurrence is recorded and mapped. It is not easy to do when there are simply two points of observation, at the beginning and end of the time of observation, such as when two aerial photos are available.

Examples of the proportion method in use
Several different variations of the analysis of proportion of the landscape disturbed over a period of observation have been done:

1. Proportion of survey lines recorded as disturbed by presettlement land surveyors. This method has been used for several studies of natural disturbance frequency in the northeastern United States. The exact time of observation is unknown, so it is necessary to bracket these analyses with minimum and maximum estimates of the length of time after disturbance that the surveyors reasonably would have recognized disturbances. Generally, authors have assumed this bracket to be equivalent to the stand initiation stage of development, lasting between 10 and 20 years. Lorimer (1977) and Canham and Loucks (1984) used land survey records and estimated rotation periods for wind of 1000–2000 years in hemlock–hardwood forest. The proportion of section corners recorded as windfall or recent burn can also be used in addition to the proportion of survey lines.

2. Proportion of the landscape occupied by early successional species. This method will work only if there is an early-successional species that can be reliably associated with recovery from stand-replacing disturbance and the average number of years this species persists among all such stands is known. Lorimer (1981) used this method to estimate fire rotation periods in a mixed forest of beech, maple, hemlock, pine and spruce in Maine. He assumed that all severe fires are recolonized by paper birch and that these stands remain dominated by paper birch

for about 75 years. Thus, the proportion of the forest dominated by birch, plus that recently burned, as recorded by presettlement land surveyors and divided into 75 years, yielded an estimated rotation of 800 years for fires in the mixed hardwood–conifer forests of north-eastern Maine.

3. Sequential air photos and/or forest inventories. The proportion of area burned or blown down on photos, or the proportion of inventory plots that are disturbed since the last inventory, are used for the analysis.

4. Stand-history reconstruction. If the disturbance history of many plots is reconstructed, one can look at the proportion of all stands that experienced a certain type of disturbance over the period of reconstruction. If the study plots have variable lengths of stand-history reconstruction then the time of observation for all plots can be pooled for a 'pseudo period' of observation. To estimate the rotation period that pseudo-period of observation is divided by the number of disturbance events that occurred among all plots. This is really the same as the proportion method because all the plots experienced at least partly the same time window and are not independent. Frelich and Lorimer (1991a) used this method and confirmed Canham and Loucks' (1984) very long rotation periods for heavy windthrow in hemlock–hardwood forest.

Confidence limits for proportion-based estimates of rotation period
When a point-sample is used to estimate rotation periods (this assumes that one is willing to consider a study plot a point relative to the landscape), the following formula will show the approximate degree of sampling error (Snedecor and Cochran 1980):

The 95% confidence interval for a proportion:

$$p \pm 1.96 \sqrt{\frac{p(1-p)}{N}} \tag{5.6}$$

where:

p is the proportion of points disturbed during some period of observation

N is the number of sample points, in this case stands that are assumed to be independent with respect to stand age.

One then applies the upper and lower bounds of the confidence interval to formula 5.5 (T/P) to get upper and lower bounds for the rotation period.

A large number of plots is necessary to get a reasonably small confidence interval for the proportion of plots disturbed over a period of observation, because the confidence intervals get narrow in proportion to the square root of the number of plots. Several hundred plots would be ideal to get a relatively small confidence interval. When working in small forest remnants, however, establishing more plots does not always help. As plot number increases the average distance between plots becomes smaller, the covariance among plots becomes larger, so that one disturbance affects many plots, and the plots are not independent. This is especially true for heavy disturbances that remove large portions of the canopy, which also tend to be geographically large (e.g. Figure 5.3). If disturbances are large enough to span the distance between sample points, then the covariance among sample points must be taken into account.

An example of this covariance occurred during Frelich and Lorimer's (1991a) study of spatial patterns in the hemlock–hardwood forest of the Porcupine Mountains. Sample stands were 1.7 km apart on average and the disturbance size distribution (Figure 2.2) was such that many disturbances were larger than 1.7 km in length. Covariance among plots was 3.4 times the variance on their study area. The confidence interval for rotation period of windstorms that remove ≥60% of the canopy ran from 805 to 9553 years (mean was 1484 years, $n = 70$) without using the covariance, but was 535 to infinity when covariance was taken into account. To bring the covariance among plots down to nearly zero the study areas would have to have been so large that only one of the 70 plots would, on the average, have landed within the larger patches created by windthrow. The mid-point of the largest size class in Figure 2.2 (6310 ha) provides a reasonable estimate of the largest patches created by windthrow during presettlement times in the Great Lakes Region. If Frelich and Lorimer had allocated one plot to each 6310 ha of forest in order to have independent samples, then the existing primary forest study areas would have had to be 19 times their current size to accommodate 70 plots. Clearly, that would be an impossible situation since one cannot create new forest for a study. This is one of those cases where random versus systematic sampling of the landscape would have made no difference: Frelich and Lorimer used a random sampling strategy, but still ended up without independent plots. Fortunately several studies, using a variety of methods, with a variety of time frames, indicate that the rotation period for heavy

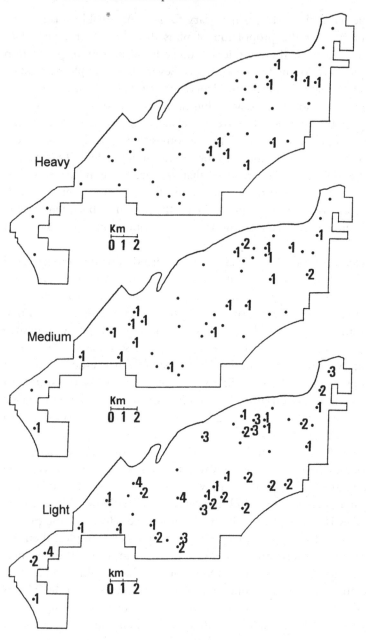

Figure 5.3. Number of heavy (>40% canopy removal), medium (20–39% canopy removal) and light (10–19% canopy removal) disturbances between 1860 and 1969 among 46 plots within the never-logged hemlock–hardwood forest of the Porcupine Mountains, Upper Michigan. After Frelich and Lorimer (1991a).

canopy destruction is in the range of 1000–2000 years. Therefore, a consensus among many studies is more useful for final inference about the frequency of stand-leveling blowdowns than the confidence interval statistics.

Confidence interval statistics are much more meaningful for smaller partial disturbances that remove 10–40% of the canopy. There is a relatively large number of observations for these disturbances and they are typically spread randomly over the landscape (Figure 5.3), so that little covariance exists among plots (Frelich and Lorimer 1991a). The rotation periods and 95% confidence limits found for partial disturbances were 52–119 years, 251–516 years, and 512–2113 years, for wind disturbances that remove 10–19.9%, 20–29.9%, and 30–39.9% of the forest canopy, respectively. Similar statistics for all disturbances removing ≥10% and ≥20% of the canopy were 40–76 years, and 98–412 years, respectively (Frelich and Lorimer 1991a).

Ecosystem types, disturbance regimes, and landscape patterns

The study of age structure and rotation periods alone does not give us a complete picture of disturbance dynamics on the landscape. Somehow, it is necessary to integrate age structure with successional stage, developmental stage, and the mosaic of ecosystem types across the landscape. Some disturbances homogenize forest type across ecosystem types. For example, intense fire can create birch forest on soils ranging from sand to silt. The absence of severe disturbance in turn may allow alternate communities to develop on different ecosystem or soil types, as the birch forest succeeds to pine on sandy soils, and maple on silty soils. Therefore, the pattern of different stand types across the landscape results from a blend of ecosystem – or soil and physiographic effects – and disturbance effects.

How do we go about characterizing this blend of ecosystem and disturbance effects? Although we cannot usually (and certainly not always) predict exactly which stands will be hit by natural disturbances, it is possible to assign probabilities that individual stands will be disturbed. For a given ecosystem type it is possible to assign one of the theoretical landscape age structures discussed earlier in this chapter. Furthermore, with enough study it is possible to link landscape age structure and the estimated proportion of stands in each age class to developmental and successional stages, using the integrating concept of the vegetation growth stage (see 'What is a vegetation growth stage?' below). It is then possible to analyze how changes in disturbance frequency will affect the proportion of stands in each vegetation growth stage on a given ecosystem type. We need only one more step to characterize a landscape that has a mosaic of ecosystem types. This is to repeat the vegetation growth stage exercise for each ecosystem type so that we have a complete picture of the expected number of stands in each vegetation growth stage on every forested ecosystem type across the landscape.

What is a vegetation growth stage?

Basically, vegetation growth stages are an integration of the developmental and successional stages. Changes in development and composition are taken into account simultaneously so that there is a new vegetation growth stage every time the developmental stage or the successional stage changes. A simple hypothetical example of forest response to stand-replacing disturbance will illustrate this:

Developmental stages:

- stand age 0–10: initiation
- stand age 11–50: stem exclusion
- stand age 51–80: demographic transition
- stand age ≥81: multi-aged

Successional stages:

- stand age 0–40: aspen
- stand age 41–80: aspen with fir understory
- stand age 81–100: mixed aspen and fir
- stand age ≥101: fir

Vegetation growth stages:

- stand age 0–10: aspen-dominated initiation
- stand age 11–40: aspen-dominated stem exclusion
- stand age 41–50: aspen–fir stem exclusion
- stand age 51–80: aspen–fir demographic transition
- stand age 81–100: multi-aged aspen–fir
- stand age ≥101: multi-aged fir

Note that vegetation growth stages provide a more detailed breakdown of forest change over time, with more categories than either developmental or successional stages. The stages can be rather subjective as well; i.e. one may count aspen with understory of fir separately, or wait until fir enters the canopy before defining a separate successional stage or vegetation growth stage. A final comment: the commonly recognized size class systems such as seedling, sapling, pole, mature, and large tree-dominated stands may be used in place of developmental stages.

A simulation of landscape characteristics

Nothing shows scientists whether they really understand all the parts of a system as well as an attempt to develop a simulation. The key here is to estimate the proportion of the landscape in each vegetation growth stage.

Simulation structure

A typical box and arrow type of simulation run on a one-year time step is adequate for this purpose:

1. Each vegetation growth stage is represented by a box that keeps track of the proportion of all vegetation within an ecosystem within that vegetation growth stage.
2. Devise a transition matrix for annual transfers among the vegetation growth stages, as indicated by research on rotation periods and succession. The transition matrix specifies that:

- 2A. A fraction of the value in the box will be transferred to the next higher vegetation growth stage, on an annual basis, according to this formula: $1/x$, where x is the number of years a stand would spend in that vegetation growth stage in the absence of stand-killing disturbance.
- 2B. A fraction of the value in each box will be transferred to the first (or post stand-killing disturbance) vegetation growth stage, on an annual basis, according to this formula: $1/y$, where y is the length of the rotation period for stand-killing disturbance. Be careful: this assumes a constant hazard function for the duration that each stand is in a given vegetation growth stage. Adjustments to account for non-constant hazard can be made by changing the fraction disturbed annually in successive growth stages.
- 2C. If there is more than one type of stand-killing disturbance, then step 2B will be repeated using the appropriate rotation period for each additional disturbance type. Note that the transfer may not always be to

the same box for each disturbance type. There may need to be two or more post-disturbance boxes. For example, a stand of a shade-tolerant, late-successional species such as balsam fir may blow down in a heavy windstorm but remain dominated by balsam fir, even though now dominated by small seedlings of fir. If the same fir stand burned, fir may be removed and succeed to aspen. In a case such as this there would be a post-disturbance vegetation growth stage called something like 'seed-ling balsam fir' for post-wind stands, and one called something like 'seedling aspen' for post-fire stands.

- 2D. Partial disturbances (non-stand-replacing disturbances) that affect vegetation growth stage can be included also, as long as the rotation period for them is known. Surface fires, for example, may kill a fir understory beneath mature pines, thereby delaying succession from pine to fir.

3. The simulation can start with an equal proportion of stands in each vegetation growth stage and then be run until the proportions come to an equilibrium. This usually means running the simulation for at least one rotation period for the dominant disturbance type.
4. Repeat the whole exercise (steps 1–3) for each ecosystem type on the landscape, and then sum the results to see what the whole landscape would look like under a given disturbance regime.

Preparing for the simulation

A unique web of different vegetation growth stages with many different types of transfers among boxes may be developed for each ecosystem type. It is up to the investigator to have a thorough understanding of stand development and successional relationships to disturbance for all disturbance types that may occur in the ecosystem, in order to properly structure the simulation. A three-step process is necessary to work out the web of growth stages and get ready to run the simulation.

Preparation Step 1. Vegetation growth stages must be quantitatively related to age classes after disturbances of all types that will appear in the simulation (necessary to get the variable x in simulation step 2A above). The age of transition may not be accurate for every stand across the landscape, but what is needed is the average age at which stands switch from one vegetation growth stage to another. This can often be done with size/age class data from field plots, or even from forest inventory data.

Preparation Step 2. Relate the vegetation growth stages to each other with a set of 'successional guidelines' as shown in the summary of Chapter 4.

Preparation Step 3. Estimate rotation periods for the ecosystem and time period of interest (necessary to get the variable y in simulation step 2B above). Many managers of reserves in North America today are interested in the so-called pre-European settlement disturbance regime (usually the historical period 1600–1900) since their goal is to compare the current forest with what is thought to be natural.

Preparation Step 4. Decide on the most appropriate model for the theoretical age distribution (i.e. hazard function) for each ecosystem type and disturbance type.

Two words of caution about the simulation

First, those who desire high precision for the proportions in various growth stages may keep track of each vegetation growth stage as a vector containing the proportion in each annual age class, and adjust the hazard function annually, instead of averaging these over all years within a vegetation growth stage. My experience is that this increases the amount of work by quite a lot and may or may not result in different results from using each vegetation growth stage as a single pool. Given the wide variation in confidence intervals for rotation periods and the fact that they fluctuate over time, small differences among different simulation methods may not have much meaning. In fact, I recommend running all of the simulations twice, using the upper and lower bounds of confidence estimates for the rotation periods. There is no point to saying that a given vegetation growth stage, for example mature white pine, should occupy 23.57% of the landscape under a certain disturbance regime. It is more reasonable to say it would occupy 15–30%.

Secondly, I caution the reader that even the estimates of proportion of stands across the landscape using the bounds of the confidence limits for the rotation period still assume an infinitely large landscape. Because disturbances can be large (fires and derechos that affect 100000 ha are common), variation in the proportion of the landscape in a given vegetation growth stage will essentially be zero to 100% for small landscapes in one patch. If the landscape is a mosaic of ecosystem types, and each ecosystem type has many widely scattered occurrences, then the proportion in each vegetation growth stage for each ecosystem type may be quite stable (see Chapter 8 'Stability of age structure', and Table 8.1).

An example simulation – the Minnesota North Shore of Lake Superior

Simulation set-up

Land surveyor records indicate the presettlement landscape comprises three major ecosystem types: a hardwood ecosystem dominated by sugar maple stands, a white pine ecosystem, and a near-boreal ecosystem with paper birch, aspen, black spruce and balsam fir.

The three forest types have the following set-up as described under 'preparing for the simulation' above. For fire rotation periods in each ecosystem type I determined the 'best estimate' from the literature and then made that the center of lower and upper bounds with a two-fold difference. Severe fires were simulated with rotation period brackets of 100–200 years, 150–300 years, and 2000–4000 years for the spruce–fir–birch, white pine and hardwood ecosystems, respectively (data from Swain 1973, Heinselman 1973, 1981a, Frelich 1992, Johnson 1992). Surface fires were simulated with a rotation period of 40 years in the white pine-dominated vegetation growth stages in the white pine ecosystem (Heinselman 1973, 1981a). Stand-leveling wind was simulated with rotation period brackets of 1000–2000 years for all forest types (Canham and Loucks 1984, Whitney 1986, Frelich and Lorimer 1991a). Theoretical age distributions were negative exponential for stand-killing fire and combination with waiting period of 50 years for stand-leveling wind.

The following set of successional guidelines was used to set up the successional web, along with its transition probabilities from one vegetation growth stage to another for each disturbance type.

Successional guidelines for the case study

Spruce–fir birch

Vegetation growth stages (VGS) and successional guidelines:

1. Sapling stands of birch and aspen 0–10 yr after fire in any VGS
2. Pole–mature stands of birch–aspen 11–50 yr after fire in any VGS
3. Mature stands of birch and aspen with conifer understory 51–80 yr after fire in any VGS
4. Multi-aged stands of conifers ≥81 yr after fire in any VGS, or ≥81 yr after wind in VGS 3, 4, 5 or 6
5. Sapling–pole stands of conifer 0–50 yr after wind in VGS 3, 4 or 6
6. Pole–mature stands of conifer 51–80 yr after wind in VGS 4

White pine
Vegetation growth stages and successional guidelines:
1. Sapling stands of birch 0–10 yr after severe fire in any VGS
2. Pole–mature stands of birch 11–50 yr after severe fire in any VGS
3. Mature stands of birch with white pine understory 50–80 yr after severe fire in any VGS
4. Mature stands of white pine 81–120 yr after severe fire in any VGS
5. Multi-aged stand of white pine, spruce–fir 121–200 yr after severe fire in any VGS or after lack of surface fire in VGS 9
6. Multi-aged spruce and fir ≥201 yr after fire
7. Sapling–pole pine for 0–50 yr after wind in VGS 3 or 4
8. Sapling–pole spruce–fir after 0–50 yr wind in VGS 5
9. Multi-aged white pine after surface fire in VGS 5

Hardwood
Vegetation growth stages and successional guidelines:
1. Sapling stands of birch 0–10 yr after severe fire in any VGS
2. Pole–mature stands of birch 11–50 yr after severe fire in any VGS
3. Mature stands of birch with maple understory 51–100 yr after severe fire in any VGS
4. Mature stands of maple 101–150 yr after severe fire in any VGS
5. Multi-aged stands of maple >150 yr after fire in any VGS, or >120 yr after windthrow in VGS 3, 4 or 5
6. Sapling stands of maple 0–10 yr after windthrow in VGS 3, 4, or 5
7. Pole–mature stands of maple 11–120 yr after windthrow

Simulation results

There are two ways to look at the results: individual ecosystems and the whole landscape with ecosystems integrated. The individual ecosystems with successional birch forests and older multi-aged conifer forests show a lot of sensitivity to changes in rotation period. For example, the multi-aged conifer VGS for the spruce–fir–birch ecosystem ranges from 47% to 66% for the short and long estimates of rotation period. Similarly the multi-aged spruce–fir in the white pine ecosystem varies from 24% to 44% of the landscape (Table 5.1). Thus, because we cannot pinpoint natural rotation periods more accurately than two-fold brackets, we cannot predict the proportion of the ecosystem in young birch forest

Table 5.1. *Results of North Shore simulation example by ecosystem type*

Ecosystem/vegetation growth stage	Age (years)	Range for proportion of stands in each ecosystem (%)
White pine ecosystem		
Sapling birch	0–10	3.2–6.3
Pole–mature birch	11–50	11.3–19.8
Mature birch–pine	51–80	9.7–12.2
Mature pine	81–120	9.2–13.1
Sapling–pole pine	0–50	0.6–1.3
Multi-aged pine	≥121	9.9–10.7
Multi-aged pine–spruce–fir	121–200	11.8–12.4
Multi-aged spruce–fir	≥201	23.5–44.3
Sapling–pole spruce fir	0–50	1.2–1.4
Spruce–fir–birch ecosystem		
Sapling birch	0–10	4.8–9.2
Pole–mature birch	11–50	15.9–26.1
Mature birch–fir–spruce	51–80	10.3–14.9
Multi-aged conifer	≥81	46.8–66.6
Sapling–pole conifer	0–50	1.6–2.1
Pole–mature conifer	51–80	0.1–0.8
Hardwood ecosystem		
Sapling birch	0–10	0.2–0.5
Pole–mature birch	11–50	1.0–1.9
Mature birch–maple	51–100	1.0–1.8
Mature maple	101–150	1.2–2.2
Multi-aged maple	≥151	83.5–91.2
Sapling maple	0–10	0.5–0.9
Pole–mature maple	11–120	5.0–9.1

versus old multi-aged conifer forest very accurately. In contrast, the proportion of the landscape covered by multi-aged maple in the hardwood ecosystem varies only slightly as the rotation periods change (Table 5.1). This insensitivity arises because the recovery to the multi-aged condition is relatively short compared with the rotation periods for wind, which is the dominant form of disturbance in the hardwood ecosystem.

Looking at the integrated landscape with all three ecosystem types, we see that there are young birch stands in all three ecosystems, but that at later developmental stages these diverge into different forest types. There are also spruce–fir stands that originate as successional replacements to birch or to white pine in two different ecosystems. At this point the North Shore ecosystems are not fully mapped, so we cannot yet weight each of the VGSs by the relative area of the ecosystem in which they occur to get an estimate of percentage of the total landscape they occupy.

The estimates of proportion of the landscape in each vegetation growth stage could be used as a coarse-filter blueprint for a forest restoration project, or for an ecosystem management project where commercial forest is to be managed so that the landscape has natural characteristics. One of the advantages of this method is that a combination of data such as presettlement land surveys, and modern estimates of number of fires per million hectares, and risk of tornadoes and severe thunderstorm winds, combined with models of forest response to disturbance deduced from the species characteristics, could in theory be used to construct a blueprint even in regions where no natural forest remnants exist.

Old growth on the landscape

Few issues have commanded more attention from environmentalists and forest managers in recent years than old-growth forest. No two people seem to agree on exactly what old growth is. The lower the degree of human influence and the older the trees in a given stand, however, the greater the proportion of people who agree that a given stand is old growth. Also, the more severely humans impact a stand, the longer it takes to get back to the nebulous old-growth condition. That is, a plowed field would take centuries to return to old-growth conditions, whereas a stand that had partial thinnings may only take a few decades before even an old-growth purist could no longer tell that there was direct human influence. This last point assumes that one's philosophy even allows for recovery to old growth after human conversion to some other use. I don't provide a solution to this definition dilemma. Instead, I present a taxonomy of types within the general category 'old forest', and show how they relate to the landscape and the disturbance regime. The complex political and management issues surrounding old growth are probably best understood when one can see the complex web of different types of old forest and how they relate to one another.

The following definitions are sure also to stir up many ecologists who may object to Clements' terms climax and subclimax and my terms short-lived and long-lived. If one goes to Clements' papers (e.g. Clements 1936) a very simple and useful definition of climax is given as any forest capable of reproducing its own composition in the absence of severe disturbance, and subclimax as any forest not occupied by climax species due to recovery from severe disturbance. Examples from among those forests I have described in this book and that are also mentioned by Clements include spruce forest as a climax type, and jack pine as a sub-climax in response to severe fire in the boreal forest. Some people object to having old-growth forest composed of short-lived species, such as balsam fir, with an average age at death of about 40 years in Minnesota. As long as the fir replaces itself and there has not been a stand-replacing disturbance for a long time, however, the term old growth may be relevant. This is despite the fact the individual trees are not old from a human point of view: we tend to call trees older than the maximum human life-span 'old'. From a forest function point of view, why should the situation be classified differently if a fir stand has continuous replacement of trees with maximum age 40, with 200 year intervals between fires, than a hemlock stand where the trees die at age 400, and the interval between fires is 2000 years? It is all a matter of relative time scales.

A taxonomy of old forests

There are different types of old forest, such as primary forest that has never been logged, secondary old growth that was logged long ago, and forests of seral species like birch that will never become what most people call true old growth, but nevertheless are older than average for the species. It is best to refer to this as the 'dynamic web of old forest', because there is not a linear pathway or relationship among different categories of old forest.

Old forest definitions

Old forest. An umbrella term including all forests listed below.

Old-growth forest vis-à-vis Frelich. Any forest at or beyond a politically established threshold stand age, threshold diameter for canopy trees, or threshold vegetation growth stage. In Minnesota, age 120 years is commonly accepted as a threshold age for old growth.

Old-growth forest vis-à-vis Oliver. A forest stand in the last of the four basic stages of stand development. See Chapter 4 and Oliver (1981).

Primary forest or natural forest. Any forest that has never had significant human disturbance, usually taken to mean never logged. This may include young forest after natural disturbance, such as young post-fire jack pine in the near-boreal forest.

Primary old growth, or natural old growth. A subcategory of old growth that has had little or no human disturbance, or conversely a sub-category of primary forest that is in a certain stage of development beyond an established age or developmental threshold. An example would be sugar maple forest >120 years old that has never been logged.

Secondary old growth. A sub-category of old growth comprising stands that were previously logged, or had other major human disturbance that precludes them from being primary old growth. This forest may be managed for timber production. If managed for timber production, then it also falls under extended rotation forest. An example would be sugar maple that was clear cut long ago, and is now >120 years old.

Climax old growth. Forests that meet an old-growth threshold and are composed of shade-tolerant species capable of reproducing under their own canopy. Sugar maple, hemlock, and spruce–fir–cedar forests may fall in this category.

Sub-climax old growth. Forests that meet an old-growth threshold and are composed of long-lived species that are shade-intolerant. These forests may succeed to climax old growth in the absence of major disturbance. Old jack pine forests in the BWCAW, and old oak and white pine forests are examples.

Sub-climax old forest. Forest composed of relatively short-lived, shade-intolerant tree species that have reached an age beyond optimum vigor. This could include paper birch, aspen, balsam poplar, or jack pine types (in more southerly areas than the BWCAW where their lifespans are relatively short) that is beyond a threshold age adjusted to their lifespan (perhaps 70 years or so).

Extended rotation forest. Any forest beyond recommended rotation age and managed for timber production.

The dynamic interactions among old forest types

Primary forest may contain stands of sub-climax old forest, primary old growth, sub-climax old growth, climax old growth, or old growth vis-à-vis Frelich or Oliver. Perhaps this explains why many debates about old growth proceed in confusion and none of the parties involved knows what the others are talking about. To be clear, one must answer four questions: (1) Is a stand primary or secondary? (2) Is it composed of sub-climax or climax species? (3) Is it composed of short-lived or long-lived species? (4) What is the stand's age or stage of development? A key to classification of old forests for the major forest types discussed throughout this book helps clarify these questions (Figure 5.4). This static classification,

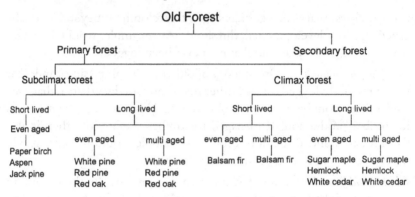

Figure 5.4. A dichotomous key to old forest types discussed in text. Only the primary-forest branch of the key is filled out – the branching structure of secondary-forest part of the key would be similar.

however, does not take into account that many of the sub-climax forest types could succeed to a climax type (e.g. birch, jack pine or aspen to fir, spruce and cedar in the colder northern parts of the Lake States, or birch to sugar maple in the warmer parts). Presumably a given stand would switch from short-lived to long-lived or from sub-climax to climax classifications when greater than 50% of the trees or basal area became long-lived or climax species. There are different degrees of movability from one branch of the key to another at different levels. For example, primary forest can become secondary forest after human disturbance, but secondary forest can never go back to primary forest. Succession can almost always carry a sub-climax type (except on very severe sites) to one of the climax types in the absence of severe disturbance. Short-lived and long-lived species can readily replace each other after disturbance or through succession (Table 5.2).

Another aspect of the old-growth issue is the size of trees. The reader may have noticed that nowhere in my definitions did I mention tree size. Contrary to the beliefs of many ecologists, I don't believe tree size should be a consideration with regard to old forest or old growth. Old forests exist with many sizes of trees. Dwarf forests are common on peatlands, sand dunes and rocky landscapes. Examples include dwarf black spruce on peatlands in northern Minnesota, dwarf oak on ridge tops in the Porcupine Mountains and on sand dunes around the Great Lakes, dwarf pitch pine on mountain tops in Massachusetts, and the ancient white cedars on rock outcrops in northern Minnesota's BWCAW. The definitions and concepts I have talked about, based on stage of development

Table 5.2. *Common successional trends in characteristics of dominant species among old forests*

Absence of stand-killing disturbance
Subclimax → climax
Short lived → short lived or long lived
Long lived → short lived or long lived
Even aged → multi-aged

After stand-killing or stand-leveling disturbance
Primary → secondary (human disturbance)
Climax → climax or subclimax
Short lived → short lived or long lived
Long lived → long lived or short lived
Multi-aged → even aged

and age structure, work equally well for these dwarf forests as for fully statured forests. To exclude such forests from consideration as old growth because the trees are shorter than humans is just as arbitrary as excluding forests dominated by tree species that do not live longer than humans. Those involved with inventory of old forest would do well to recognize old forest in dwarf, mid-statured, and fully statured categories.

Extended rotation forest (ERF) as a general category could include stands of any of the other old forest categories. ERF can fulfill some of the functional roles of primary old growth, climax old growth or old growth vis-à-vis Frelich or Oliver, while still remaining in the commercial forest base. When a landscape is managed with a proportion in typical short-rotation forestry, and some in ERF, the landscape age distribution – at least at the time the forest reaches a regulated condition – changes from a uniform distribution to a combination distribution, with the ERF in the negative exponential right tail of the distribution. Such a mixed management regime can allow a landscape age distribution that is closer to a natural age distribution than a uniform age distribution.

When the definition for old growth consists of a simple age threshold, then it is relatively simple to calculate how much old growth there would be, as long as one knows which of the theoretical age distributions applies for a given forest type. The proportion of old growth stands is the integral of area under the curve beyond the threshold age (Figure 5.1). When old growth definitions are more complicated the simulation methods used earlier in this chapter (see 'Ecosystem type, disturbance regimes and landscape patterns') are appropriate for estimating the amount of old growth that could exist under a given disturbance regime. For example, the multi-aged white pine–spruce–fir and multi-aged white pine within the white pine ecosystem (Table 5.1) could be considered old growth by Oliver's definition.

Summary

Chapter 5 has examined how to recognize major types of landscape age distributions and how to infer rotation periods from these distributions. The uniform age distribution can only occur in fully managed forests where the maximum stand age is equal to the rotation period (i.e. all stands are harvested at a given age, and a constant proportion of all stands is harvested annually). The negative exponential age distribution results from natural disturbances that can replace stands of any age so that the hazard of disturbance is constant with stand age. The rotation period is equal to the mean stand age. Crown fires in northern closed-canopy forests such as jack pine fall into this category, although one needs to recognize that disturbance frequencies change over the centuries, often resulting in mixed age distributions. The Weibull distribution is best used to model age distributions in forest types with a constantly increasing hazard of disturbance with stand age. Finally, in some forests the shape of the age distribution changes from one model to another at a threshold age. The best example in the Lake States is sugar maple–hemlock forest – in which wind is the primary disturbance type – and stands are not susceptible to windthrow when young and are then subject to random windthrow after age 100–130. This results in a shift from the uniform age distribution model for young stands to the negative exponential model for older stands. This results in a true compound distribution of different models, not simply a shift in rate of disturbance over time that causes a mixed distribution with different slopes for the same model.

The concept of the vegetation growth stage (VGS) was introduced. The VGS integrates developmental (or structural) changes and successional (or compositional) changes in stands over time. A method was given to estimate the proportion of stands across the landscape that would fall within each VGS for a given disturbance regime, using the earlier information on stand age distributions and rotation periods. Rotation periods vary a lot and it is best to use two-fold or more brackets for rotation periods to estimate a reasonable range of variability for the proportion of the landscape occupied by each VGS. Such simulated estimates can be used as a rough blueprint for restoration work in forests or for comparison of the managed landscape with a natural landscape.

Finally, this chapter considered the complex issues surrounding old growth or old forest. One needs to consider the degree to which a given forest has been influenced by people, the successional status of the species, how long-lived the species are, and the stage of development, which may roughly correspond to age of stand. In other words, is a stand primary or secondary, is it in climax or subclimax condition, is it dominated by trees that replace each other on short intervals or trees that live for centuries, and is it in initiation, stem exclusion, demographic transition, or multi-aged stages of development? There is no way to clearly discuss the political issues surrounding old growth without knowing these factors. In addition, old forests may be dwarf, middle, or fully statured. Dwarf forests are likely to occur on peatlands, rock outcrops, and sand dunes; middle-statured forests occur on moderately good soils; and fully statured forests occur on sheltered sites with deep, high-quality soil. Forests growing under these three different conditions are likely to have different ecosystem function and associated species even when dominated by the same tree species. Therefore, forests of different stature should not be ignored by those involved with old forest or biodiversity issues.

6 · The disturbance regime and landscape structure

Knowledge of stand dynamics (Chapter 4) and landscape structure (Chapter 5) are combined in this chapter to examine in detail the relationship between disturbance and landscape characteristics. The topics progress from wind regimes to fire regimes to complex regimes with wind, fire and herbivory (this latter topic was introduced in Chapter 2). Detailed case studies of landscape characteristics and their sensitivity to changes in disturbance regime are considered in each case. Then two more important issues are introduced that can interact with fire, wind and herbivory to provide even more complexity to landscape structure: (1) disturbance size, and (2) interactions among trees themselves, or neighborhood effects. As we will see here and Chapter 8, trees superimpose a patch-forming mechanism of their own onto that created by disturbance dynamics.

Wind regimes and landscape structure

Windstorms have a great effect on structure and development of individual stands across the landscape in the Great Lakes Region. They pockmark the landscape with young stands in initiation and stem-exclusion phases of development. If wind is the dominant disturbance type in an ecosystem, the combination distribution of stand ages will result, because stands are not susceptible to massive blowdown until they reach the late stem-exclusion or demographic-transition phases, after which they will be hit at random by stand-leveling windstorms. Winds of less than stand-leveling force also create gaps in older stands, thereby determining the timing and size of new cohorts in multi-aged forests. Thus, the wind regime largely determines the proportion of stands on the landscape in each of the eight stages of development described for hemlock–hardwood forests in Chapter 4. One way of illustrating how wind regimes influence the landscape is to look at the sensitivity of landscape structure, in terms of proportions of stands in various developmental stages, to changes in the windstorm frequency. The STORM simulation (Frelich and Lorimer 1991b) was developed to look at stand structural changes associated with complex wind regimes, and will be used for the sensitivity analysis.

Simulation structure

The simulation is similar to that described in Chapter 5, although stand structure is simulated in more detail, and it runs on a 10-year (rather than 1-year) time step. At each time step there is a suite of probabilities for wind disturbances that remove anywhere from 10% of the canopy area (defined as exposed crown area from Chapter 3) to ≥70% of the canopy. If there is a disturbance during a given decade, then a portion of the canopy is removed, and a new cohort enters the forest. All cohorts are assumed to have a normal diameter distribution to start with. In the absence of disturbance the mean diameter and variance increase over time in accordance with empirical data on mean/variance of tree diameter by age class from the Porcupine Mountains study area in Upper Michigan (Figure 6.1). The proportion of each cohort in four standard

Figure 6.1. The never-logged hemlock–hardwood landscape of the Porcupine Mountains: a rare example of a fairly natural landscape with all stages of development present. Photo: Lee Frelich.

size classes (sapling, ≤10.9 cm dbh; pole, 11.0–25.9 cm dbh; mature, 26.0–45.9 cm dbh; and large, ≥46 cm dbh) is kept track of over time. When a disturbance occurs, the portion of the canopy removed is distributed so that 1.5 times as much canopy area is removed from large trees as from mature trees, and 1.5 times as much from mature trees as from pole trees, as indicated by empirical evidence on relative susceptibility of trees to windfall (Frelich and Lorimer 1991b). For example, a disturbance removing 20% of the forest canopy in the entire stand would take slightly more than 9% of total plot crown area away from the large trees, about 6% out of mature trees, and about 4% out of pole-sized trees. These portions of canopy area are transferred to the sapling size class as a new cohort. Since individual cohorts – especially older ones – often have trees scattered among pole, mature and large size classes, they lose their initial normal dbh-distribution shape as time goes by and they sustain disturbances. (See details in Frelich and Lorimer 1991b.)

To look at landscape structure with this simulation I ran the model for 1000 stands and for 1000 years to make sure any initial conditions worked their way out of the system. All 1000 stands were then classified into one of the eight developmental phases (Figure 4.8) and the proportion of the 1000 stands in each phase of development was tallied.

Sensitivity of the hemlock–hardwood landscape to differing wind regimes

Windstorm frequency varies significantly across the part of eastern North America occupied by hemlock–hardwood forest. There is a four- to five-fold variation in tornado frequency across the region (Thom 1963). Also, parts of the hemlock–hardwood forest are susceptible to tropical cyclones (hurricanes). Major hurricanes hit New England approximately every 50 years, but heavily impact only a fraction of the region each time (Boose *et al.* 1994). Between these hurricanes the severe storm regime is much more moderate than in the Lake States. To encompass this geographic range of variability in wind regimes, let us compare landscape structure under the historic disturbance regime (that in effect during 1840–1969 in Upper Michigan; Frelich and Lorimer 1991a), with landscape structure under $\frac{1}{2}\times$ and $2\times$ the historic disturbance frequency scenarios, and a New England hurricane scenario. This latter scenario features heavy windthrow (>70% canopy removal) every 250 years (the hurricanes) with light windthrow – similar to the $\frac{1}{2}\times$ simulation – between the hurricanes. The disturbance frequencies for these four scenarios are summarized in Figure 6.2.

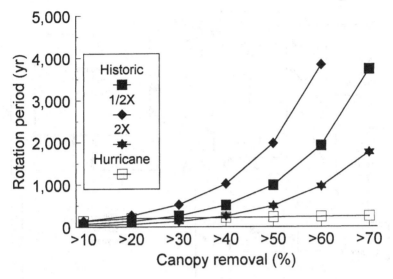

Figure 6.2. Rotation periods as a function of percent canopy removal used for historic, $\frac{1}{2}\times$, $2\times$ and hurricane simulations.

The 'historic disturbance regime' scenario shows that the old multi-aged stage of development is the hub of the whole landscape, with 50% of the landscape in that category (Figure 6.3). Switches in and out of the old multi-aged stage dominate the dynamics of the system. These include switches to pole and mature multi-aged stands after partial windthrow and subsequent recovery to old multi-aged forest. The frequency of even-aged sapling, pole and mature stands, and steady state stands is low for this scenario.

It appears that the debate over the natural condition of the hemlock–hardwood landscape – whether it was a climax forest landscape covered by steady state stands versus a landscape dominated by even-aged post-disturbance stands – that went on from the 1940s through the 1980s was overly polarized. Graham (1941b), Maissurow (1941), Raup (1957, 1981), Henry and Swan (1974), and Heinselman (1981a) emphasized major disturbances while Hough and Forbes (1943), Leak (1975), Lorimer (1977), Bormann and Likens (1979), and Lorimer (1980) emphasized climax forests, although they also recognized that major disturbances occurred. The pendulum swung from one extreme to the other during this debate and almost missed what was really happening in the forest: most disturbance that dominates the dynamics is between the two extremes. Most of the trees die during disturbances that remove

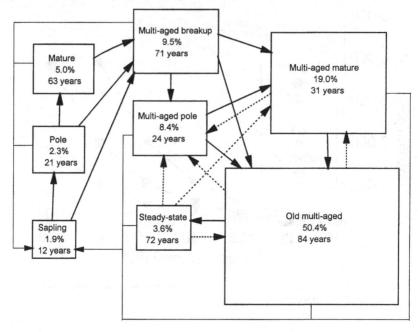

Figure 6.3. Box-and-arrow diagram of Upper Michigan, hemlock–hardwood forest landscape dynamics under the historic disturbance regime. The boxes represent eight stand types as defined in Box 4.1. The percentage of the landscape occupied by each stand type (also proportional to size of box) and average residence time for a stand in each type are shown in the boxes. The consolidated arrows with square corners represent transfers to the sapling stage after stand-leveling wind, solid single arrows represent advances in stand stature due to growth and development, and the dotted arrows represent reductions in stature caused by partial windthrow. After Frelich and Lorimer (1991b).

10–30% of the forest canopy within a given stand, and most young saplings entering the canopy do so in a neighborhood of all young trees, surrounded by neighborhoods of older trees (Frelich and Lorimer 1991a, Frelich and Graumlich 1994). Very small treefall gaps are very common but have a modest cumulative impact. Stand-leveling winds can cover large areas but are very rare, so that their cumulative impact is also modest. The moderate windstorms causing 10–30% canopy mortality have the 'winning' combination of moderate frequency and moderately large number of trees affected for each occurrence (Figure 6.4).

The $\frac{1}{2}\times$ and $2\times$ historic scenarios show that changes in wind frequency have synergistic effects: (1) doubling windstorm frequency results in four times the proportion of young post-blowdown stands on

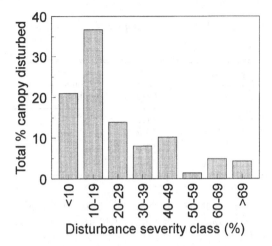

Figure 6.4. Percentage of all canopy area removed as a function of wind disturbance severity in hemlock–hardwood forest of Upper Michigan.

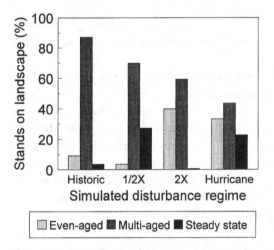

Figure 6.5. Landscape structure for the four simulated disturbance regimes in hemlock–hardwood forest. After Frelich and Lorimer (1991b).

the landscape and cuts the proportion of old multi-aged stands to one-quarter of its value in the historic simulation; and (2) cutting the wind-storm frequency in half causes slight reductions in the proportion of the landscape in even-aged types and multi-aged types but a huge increase (more than seven-fold) in the proportion of the landscape occupied by steady-state stands (Figure 6.5). The 'hurricane' scenario provides an

Table 6.1. *Comparison of tree size distribution[a] and canopy residence time for four simulated disturbance regimes*

Simulation	Size class				Average canopy residence time (yr)
	Sapling	Pole	Mature	Large	
Historic	12.9	17.7	28.2	41.2	175
½×	10.4	15.1	25.7	48.8	213
2×	21.9	27.7	32.6	17.8	111
Hurricane	15.7	19.5	29.0	35.8	161

Note:
[a] Percentage of crown area occupied by sapling (0–10.9 cm dbh), pole (11.0–25.9 cm dbh), mature (26.0–45.9 cm dbh), and large (≥46.0 cm dbh) size classes.
Source: After Frelich and Lorimer (1991b).

interesting comparison: there is almost as much even-aged types as in the 2× scenario, and almost as much steady-state forest as in the $\frac{1}{2}$× scenario. Multi-aged types are pinched out, except for multi-aged break-up stands that are apparently on their way to becoming steady state stands (Figure 6.5).

A look at the proportion of trees in the four major size classes (sapling, pole, mature, and large) for all stands pooled across the simulated landscapes shows remarkable resistance by the forested landscape (with one exception) to changes in the disturbance regime. For example, the proportion of trees in the four size classes is almost identical for the historic, $\frac{1}{2}$×, and hurricane scenarios (Table 6.1). The one exception is in the size distribution for the 2× regime, which has half the large trees and significantly more sapling and pole-sized trees than the other scenarios.

There is a damped response of average canopy residence time across the landscape to changes in windstorm frequency. Residence time rises in the $\frac{1}{2}$× scenario, and falls in the 2× scenario, but in far lower proportions than the change in windstorm frequency (Table 6.1). This damped response is due to the relative lack of susceptibility of sapling and pole trees to blowdown. After one blowdown event, it takes several decades for a new crop of susceptible trees to accumulate (Runkle 1982, Foster 1988a, b, Frelich and Lorimer 1991a). This blowdown-frequency regulation effect by the forest may explain why there is so little variability in canopy residence times among temperate zone forests (Runkle 1982).

Fire regimes and landscape structure

Fire in the hemlock–hardwood forest

Hemlock–hardwood forests were earlier classified as having fires on rare occasions (Table 2.2). Fires occur at a given point on the ground only once every several hundred years (surface fires), or 2000–3000 years (severe fires). Why then is this subsection on fire included? Why is a rare disturbance important?

The first answer to these questions is that fires, despite their rarity, play a major role in maintaining diversity. Several tree species, including paper birch, red oak, and white pine only enter hemlock–hardwood forests after fires. All three of these species take advantage of special sites scattered throughout the forest, such as rock outcrops, where they are protected from invasion and replacement by hemlock and sugar maple. It is hard to explain how the large and heavy seeds of red oak get dispersed into burned areas. Every time I encounter a red oak tree in the middle of huge tracts of sugar maple and hemlock forest, however, there are invariably fire scars on nearby trees. Apparently these small groves of red oak (Figure 6.6) get established after spot fires that burn a fraction of a

Figure 6.6.   Large red oak (left foreground) in a sugar maple forest. Photo: University of Minnesota Agricultural Experiment Station, Dave Hansen.

Table 6.2. *Rotation periods for fire and dominance or presence of paper birch,*
white pine and red oak in the hemlock–hardwood forest

Species	Rotation period for fire (yr)		
	500	1000	2000
Dominance after severe fire (proportion of stands on landscape)			
Paper birch	0.16	0.08	0.04
White pine/red oak	0.30	0.15	0.08
Presence after fire (proportion of stands on landscape)			
Paper birch	0.40	0.20	0.10
White pine/red oak	0.70	0.35	0.18

Note:
Assumes length of dominance/length of presence are 80/200 years for paper birch,
and 150/350 years for white pine and red oak.

hectare. How the acorns of red oak get into all these spots is a mystery,
although it is reasonable to hypothesize at this point that they are brought
in by birds. Bluejays, in particular, are known to carry acorns long distances (Bossema 1979, Darley-Hill and Johnson 1981). It is not known if
Bluejays behave like frugivorous birds in forests, which are attracted to
gaps in the forest and disperse plant seeds there (Hoppes 1988), or
whether they are always depositing acorns throughout the forest and only
those deposited in fire-caused gaps experience successful establishment.
The latter is certainly a reasonable hypothesis, since red oak seedlings are
much more likely to succeed on the forest floor if sugar maple seedlings
are removed by fire.

Presettlement surveys indicate that within mesic hemlock–hardwood
landscapes in the Lake States, stands with heavy dominance by paper
birch, red oak and white pine ranged from 1–2% of the landscape in
Upper Michigan to 8–10% in central Wisconsin (Bourdo 1956). Analysis
of forest remnants that have not been logged is consistent with these estimates, with about 3% of stands dominated by these species (Frelich and
Lorimer 1991a). If there was a 2000-year rotation period for fires that
allowed these species to enter hemlock–hardwood forest then paper
birch dominance, which lasts about 80 years, would occupy about 4% of
the landscape. The oldest paper birch known in Upper Michigan are
about 200 years old so that about 10% of all stands (200/2000) would
have at least a few paper birch present (Table 6.2). Similar calculations for
white pine (assuming the species can dominate a stand for 150 years and

live a maximum of 350 years) indicate that it would be dominant on about 8% of the landscape and present on 18%. Red oak may happen to be the species that invades after fire and it also has potential to dominate 8% and be present on 18% of the landscape with a 2000-year rotation period for severe fire (Table 6.2). The aforementioned spot fires could also allow presence of small numbers of paper birch, white pine and red oak in stands scattered throughout the landscape.

The second answer to the question of why rare disturbances are important is that they exhibit how the forest responds to disturbance types that are rare during one era, but may become common in a different era. We have just seen that early to mid-successional forests of paper birch, white pine and red oak occupied a small portion of the original hemlock–hardwood landscape in presettlement times. For example, the type of disturbance that allowed these species to enter the forest (burning in windfall slash) was made into the dominant disturbance type for several decades by settlers who unintentionally mimicked the natural wind–fire interaction by substituting logging for wind and setting the fires themselves in a widespread fashion. They converted much of the hemlock–hardwood landscape to birch and oak. We are still living with the forest legacy of those land clearing decades from 1840 to 1920.

Fire in the near-boreal jack pine forest

Under the natural disturbance regime these forests in northern Minnesota were equally likely to burn regardless of stand age. As we see below, this fact alone determined the structure of the landscape in terms of species composition. First, however, I will try to convince the skeptical reader that young stands really are as likely to burn as older stands. Do not mistake the statement 'equal probability of burning for all stand ages' for equal intensity of burning or equivalent fire behavior. If one goes out to a 10-year-old jack pine stand on a hot day after a prolonged period of drought and throws a lighted match in the stand, it will go up in flames. So will a 100-year-old or 200-year-old stand. They have equal probability of starting on fire but they will not burn the same way. Flame lengths in the older stands will surely be much longer, indicating a higher fire-line intensity. The 3–4 m flame lengths in the 10-year-old stand, however, are quite sufficient to kill the young jack pines with their thin bark. Many also forget that fires create huge amounts of fuel as well as consume fuel – hence the counter-intuitive constant hazard function and the myth that young forest stands are less likely to burn. A jack pine stand

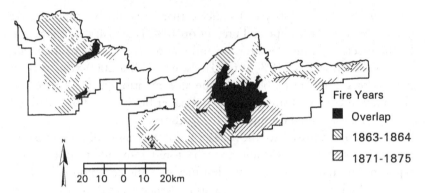

Figure 6.7. Burned twice in a short time: a large area of overlap where two stand-replacing fires occurred within 7–13 year periods within Heinselman's (1973) BWCAW study area.

that burned 10 years ago would have hundreds of dead and dried snags per hectare as remnants of the pre-fire stand and rather dry foliage with a high bulk density close to the ground – thus its likelihood to burn. These observations have also been substantiated by several field studies showing the negative exponential age distribution at four boreal or near-boreal forest sites (Van Wagner 1978, Johnson 1992), and by observations that large intense fires – those that do most of the burning across the landscape – do not stop for patches of young forest. The large zone of overlap of the 1864 and 1875 fires as reconstructed by Heinselman (1973) in northern Minnesota is a prime example (Figure 6.7).

The random timing of fires was a double-edged sword for jack pine. Most of the time fires occurred when stands were at such an age that jack pine can replace itself via massive post-fire seeding. However, some fires occurred before jack pines were old enough to bear seeds (as Figure 6.7 shows) and some fires occurred after too long, when only a few old jack pine remained in the stand (Frelich and Reich 1995a). How long does it take late-successional species to replace jack pine, and how many stands go that long without fire? Observation of jack pine stands in 1934, 1961, and 1991, showed that young stands (<40 years old) in 1934 were still dominated by jack pine as of 1991. Mature stands – those averaging 70 years old in 1934 – were almost all still dominated by jack pine in 1961, but only 70% were dominated by jack pine in 1991. Stands that were old (101–150 years) in 1934 were about 70% jack-pine dominated in 1961, but none was still dominated by jack pine as of 1991 (Figure 6.8). These

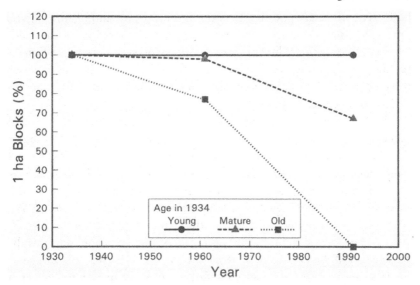

Figure 6.8. Empirical observations of loss of jack pine dominance from 1934 to 1991, depending on initial stand age. Percentages of 1-hectare blocks of forest that retained dominance by jack pine are shown, determined by analysis of sequential aerial photos. From Frelich and Reich (1995a).

data show that jack pine dominance ends somewhere in the age range of 120 to 180 years among stands.

The historic disturbance regime in the jack pine forest was one of frequent severe crown fires with a rotation period of 50–70 years (Van Wagner 1978, Heinselman 1973). Most of the stands on the landscape would be dominated by jack pine with little succession as long as fires continued to occur at intervals appropriate for jack pine reproduction (see Chapter 4 'Time lines of disturbance and stand development'). There have been major changes in fire frequency in Minnesota's near-boreal forest during the twentieth century. During that century the rotation period for fire was 500–1000 years (Heinselman 1973). The cause of this shift in rotation period can be attributed to fire exclusion, by changing the forest around the Boundary Water Canoe Area Wilderness (BWCAW) into less flammable forest types and other non-flammable land uses. Climatic change leading to less severe burning weather and direct fire suppression also played a role in reducing fire frequency. The results were: (1) more old forest on the landscape; (2) more contiguous distribution of late-successional species, resulting in spread of spruce budworm, an insect pest that lives on fir and spruce; and (3) loss of jack pine seed source due to senescence of jack pine trees before the seeds were released by fire. This last effect could change the system indefinitely because seed sources are different now than they were for the previous three centuries. Thus, fires will have a different impact on future succession than they had in the past.

The formerly predominant model of successional direction – parallel succession with jack pine replacing itself – is now changing rapidly under the present fire frequency. Many stands are now entering a successional stage that was rarely seen in the BWCAW prior to 1910. Under the historic rotation period the chance that a given point on the ground would survive 100, 150 and 200 years without stand-killing fires, was 13.5%, 5.0% and 1.8%, respectively. Currently, stands >100, >150 and >200 years old occupy about 62%, 14% and 5% of the landscape, respectively. Nearly half of the forest (about 48%) is in the 100–150 year-old category. In absence of fire this old jack pine forest will make its way to ages such that jack pine cannot replace itself within a few decades. Once jack pine seed source is lost in a given locality, reinstatement of burning may lead to aspen forests, and continued lack of fire will lead to the old growth mixture of balsam fir, black spruce, white cedar and paper birch. Both of these options lead to vegetation types that were of minor importance in the presettlement near-boreal forest and exclude the previously dominant mosaic of even-aged jack pine forests.

Complex disturbance regimes and landscape structure

A case study of white pine forest dynamics illustrates the types of effects that complex disturbance regimes can have on forest structure. Minnesota white pine forests have a complex interaction of wind, fire and herbivory by white-tailed deer and a complex successional web that governs white pine's abundance on the landscape. A frame-based model was used to find out how best to create and maintain white pine stands (Tester *et al.* 1997, Figure 6.9). A set of rules was devised to determine under what conditions white pine seedlings will become established, how fast they will grow, whether they will successfully mature and replace jack pine or aspen, and whether they will be able to keep late-successional spruce or maple at bay:

- Jack pine or aspen stands are created by crown fire or logging.
- The model will switch from the jack pine frame or aspen frame to

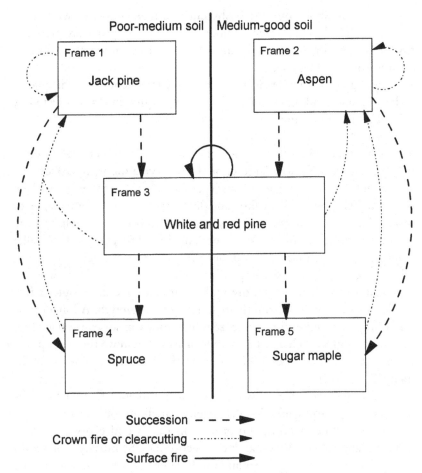

Figure 6.9. Structure of the frame-based model from Tester *et al.* (1997). Long-dashed arrows, short-dashed arrows, and solid arrows represent transfers due to succession, crown fire or clear cutting, and surface fires, respectively.

white pine if a cohort of white pine successfully makes it to the canopy by reaching 18 m in height. They must also reach 9.0 m in height to survive a surface fire.

- The model will remain in the white pine frame as long as new cohorts of white pine continue to replace old ones as they die, there is no crown fire or clear cut, and succession to maple or spruce is prevented by surface fire.
- New cohorts of white pine can enter the forest in the understory of jack pine and aspen stands. They can also enter mature white pine

stands after surface fire. Establishment of a new cohort of white pine seedlings can be suppressed if understory maple or spruce are present.

- Old cohorts of white pine can be removed by crown fire, old age, windstorm and logging.
- Height growth of trees is a species-specific function of soil quality, drought and shading by taller trees. Seedling growth is also affected by level of deer browsing.

The basic story here is that young cohorts of pine must get established in the understory and grow large enough to survive the next surface fire. This gauntlet that young pines must run may be short if there are no deer, no droughts, good soil quality and little competition from maple or spruce. The time it takes for new white pine to become established will be long if there are many deer, frequent droughts, poor soil, or taller maples and spruce in the forest understory. The longer it takes for young white pine to grow to a size large enough to survive the next surface fire, the greater the chance that a fire will occur and wipe them out, or that maple or spruce will overtop them and stop their growth. A longer time for young white pine to reach a size where they can survive surface fires also means a greater chance that windstorms will remove the older pines, leaving the stand without a seed source for future white pine establishment (Figure 6.10).

Simulations with tree growth rates and disturbance frequencies calibrated for poor soil (jack pine ecosystem), medium soils (mixed conifer and hardwood ecosystem), and good soils (hardwood ecosystem) were run for a landscape of 1000 stands each ecosystem. Sensitivity analyses for a factorial combination of deer and fire frequencies were also run (Figure 6.11).

There are three principal results: (1) optimal fire management (suppressing or preventing crown fires and using prescribed surface fire at certain times) and keeping deer density low are necessary to maintain white pine in a stand for a long time; (2) medium soils are by far the best for maintaining white pine; and (3) windstorms are by far the most common cause for loss of white pine in a given stand (Tester *et al.* 1997).

Why are medium soils best for white pine? The species actually grows faster and attains larger size on good soils. A comparison of the situation for each soil type reveals the answer. On poor soils, deer browsing in combination with nutrient-limited growth of white pine seedlings and a high frequency of surface fires means that white pine seedlings are usually

Deer browsing high   Deer browsing low

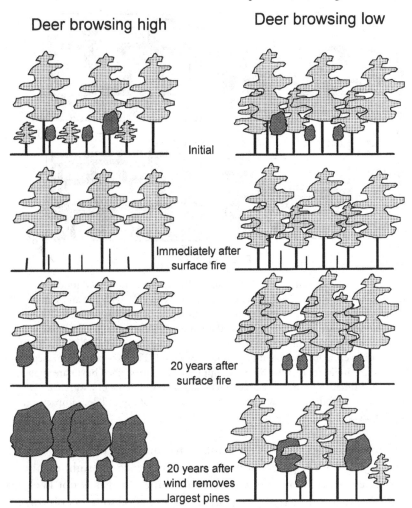

Initial

Immediately after
surface fire

20 years after
surface fire

20 years after
wind removes
largest pines

Figure 6.10. Sensitivity of forest composition to deer browsing. On the left, deer browsing prevents white pine saplings from reaching the size necessary to survive a surface fire. Many years after the fire, a windstorm removes canopy pines, leaving only maples. On the right, deer do not browse white pine saplings, so that they grow large enough to survive a surface fire, and they can move into the canopy after wind removes the oldest cohort of pine, insuring that white pine will be maintained as a component of the forest for another century or two.

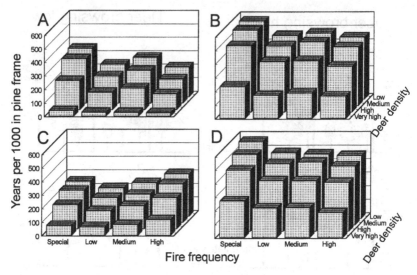

Figure 6.11. Deer and fire sensitivity analysis from the frame-based model of Tester *et al.* (1997). A, poor soil ecosystem (the BWCAW), with fire and wind disturbance types. B, medium-quality soil ecosystem (Itasca State Park, Minnesota), with fire and wind. C, good-quality soil ecosystem (Minnesota North Shore of Lake Superior), with fire and wind. D, good-soil ecosystem with a 20% chance of clearcutting each decade in addition to fire and wind. The fire-frequency category 'special fire' indicates that wildfires are suppressed and prescribed burns are applied at optimal times to favor white pine regeneration. Wind frequency was held constant for all four scenarios to facilitate comparison of fire and deer impacts on white pine.

not able to attain a size large enough to survive before the next fire occurs. Only a low proportion of stands experience the right combination of events to allow white pine establishment and it is not likely to persist there for more than one generation. On good soils, deer browsing and overtopping by shade-tolerant maples prevents pine seedlings from surviving until they are able to enter the canopy. Those that do enter the canopy have a high probability of being removed by wind, and there is usually no younger pine cohort underneath to replace them after the storm. Finally, let's consider medium-quality soils. Here white pine seedlings grow relatively fast compared with maple seedlings and surface fires are not quite as frequent as on poor soils. In the absence of high levels of deer browsing, seedlings are likely to be large enough to survive the next surface fire, while the maples are not. If windstorms remove an old cohort there is likely to be a younger white pine cohort in the stand. Neither fire nor windstorms can exterminate white pine very often, and white pine dominates longer and prospers in a much greater proportion of stands than on either good or poor soil (Figure 6.11).

White pine's chances for dominance on good soils can be enhanced markedly by occasional clear cutting of sugar maple stands followed by slash burning. The resulting aspen stands give white pine repeated chances to enter a forest type from which they normally would be excluded. Of course, this assumed an abundant seed source for white pine in the vicinity and deer control or protection of individual seedlings from browsing.

Disturbance size: when does it make a difference?

General concepts

I have already said that disturbance is a spatial process. Likewise, succession is a spatial process in addition to a temporal and directional one. There has been much debate about whether a given species is present from the start of stand initiation – what I call 'in place succession' – versus 'invasion succession' or 'wave-form succession' where late-successional species are not present at stand initiation. This latter definition of succession corresponds to that commonly used by most ecologists. If late-successional species are present from the start, then apparent succession occurs because difference among species in dominance by biomass or basal area can be due to differential growth rates. There is no definitive rule to distinguish between major fluctuation among species (see Chapter 4) versus in-place succession.

In reality, each disturbance creates many stands along a gradient of distance from refugial seed sources, or places where a species survives disturbance. In some stands late-successional and/or shade-tolerant species will be present immediately, while others may be far removed from the presence of certain species. After disturbance, species begin to spread from their refuge causing what Frelich and Reich (1995a) have termed wave-form succession. A gradient of stands will be invaded at different stand ages and stages of development by the same species merely because of the spatial layout of stands (Figure 6.12). Therefore, generalizations as to whether succession is due to invasion or differential growth are irrelevant. One just needs to characterize the wave-form succession to predict how and at what time it will affect a given stand. Wave-form succession usually occurs for any species that does not survive by some mechanism within the perimeters of a disturbed area.

Species can be ranked based on how dependent they are on refuges

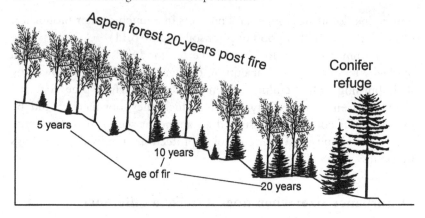

Figure 6.12. Hypothetical case of wave-form succession emanating from a shoreline refuge, as is typical in the boreal forest.

and resulting wave-form succession. Those with low recovery time to disturbance interval ratios (R:DI) will be most restricted to refuges and the immediate vicinity. Recovery time in turn depends on the rate of movement and distance between refuges. Rate of movement depends on ability to grow in the post-disturbance forest, age at which seeds are produced, and effective seed dispersal distance. R:DI ≥ 1.0 means that species can always dominate the landscape, at least for a period of time between disturbances, regardless of refuge location. R:DI much less than 1.0 indicates a species that is almost totally restricted to its refuges from disturbance.

Application of the concepts

At first thought it seems that each species should be able to be classified as a 'wave-form succession' species or an 'in-place succession' species. This is not the case, however. A look at the literature from the major forest types in the Lake States shows that the type of succession that takes place after disturbance varies from one disturbance type to another, and also from one landscape to another. There is an interesting relationship of the mechanism by which species survive disturbance and whether they will be able to have in-place or wave-form recovery (Table 6.3). The tree species in the Lake States can be classified into five categories: serotinous, sprouters, long-distance dispersers, fire resistant, and fire sensitive. These correspond to surviving disturbance in the form of canopy-stored seeds, as underground rootstocks, no survival on the disturbed site, survival as

Table 6.3. *Mechanism of survival and recovery after certain types of disturbance for key tree species in the Lake States*

Survival strategy	Crown fire	Surface fire	Stand-leveling wind	Treefall gap
In-place recovery by serotiny (canopy-stored seed)	Jack pine Black spruce	Jack pine Black spruce		
In-place recovery by sprouting (or nearly instant long-distance seed dispersal)	Red oak White oak Aspen Paper birch	Red oak White oak Pin oak	Red oak Red maple Basswood	Red oak Red maple Basswood
In-place recovery by mature tree survival		White pine Red oak White oak Sugar maple Hemlock Pin oak		
In-place recovery by seedling survival			Hemlock Sugar maple Black spruce White cedar Balsam fir	Hemlock Sugar maple Black spruce White cedar Balsam fir
Wave-form recovery by short-to-medium seed dispersal	Sugar maple Hemlock Balsam fir White cedar Yellow birch Green ash Basswood Red maple	Yellow birch Green ash Basswood Red maple	Yellow birch Green ash Basswood Red maple	Yellow birch Green ash Basswood Red maple

mature adults, and as seedlings (this latter one after disturbance other than fire), respectively. We do not seem to have any tree species with soil-stored seed banks, although there are several such herbaceous plants, including *Geranium bicknellii*, *Corydalis sempervirens*, and *Adlumia fungosa*. Those species surviving as canopy-stored seeds are most efficiently perpetuated by severe fires, those surviving as mature adults – usually thick-barked species – are ideally perpetuated by surface fire, and those species surviving in the form of seedlings are most efficiently perpetuated by windstorms that release the seedlings from suppression. Sprouters in general can get along after any type of disturbance, provided they were already present, with the exception of landscapes covered by thin rocky soils, where they may be exterminated by severe fire. Long-distance dispersers have no need to survive disturbance, although these species are also sprouters (Table 6.3). We can draw a general principle from the patterns in Table 6.3 that finally answers the question asked at the beginning of this section (Disturbance size: when does it make a difference?):

> Each species has 'in-place succession' after a certain disturbance type, presumably the one to which it has evolutionary or chance adaptation, and each species must get along by 'wave-form succession' after other types of disturbance. When there is a species-disturbance combination that works on the wave-form model, then disturbance size matters. Otherwise, disturbance size does not matter because the species survives at numerous points within the disturbance perimeter.

When in-place succession occurs by a given species, the effective disturbance size is not that of the disturbance perimeter, but rather the average distance from a surviving propagule or individual. For example, the mean distance from a random point within the perimeter of the large 1988 Yellowstone fires to a surviving tree is much smaller than the mean distance to the disturbance perimeter (Turner *et al.* 1997).

Some case studies of disturbance size and successional response

Does stand-leveling blowdown size make a difference in post-disturbance successional sequence? Analysis of canopy blowdowns in the hemlock–hardwood forest of the Porcupine Mountains shows virtually no change in composition after the disturbance for a range of sizes from 1 ha to 1000 ha (Frelich and Lorimer 1991a, Romme *et al.* 1998). Advanced regeneration that survives windthrow reflects the composition of the overstory (Frelich and Reich 1995b) and forms a new stand with the same composition after windthrow. The effective disturbance size here is zero, because within the disturbance perimeter after heavy windthrow one is never more than a few meters from a surviving seedling. Conversely, a 100-ha blowdown that burned during the 1930s in the Porcupine Mountains is covered today with paper birch and aspen forest. Sugar maple is just beginning a wave-form invasion of the understory now that the forest is 70 years old (Frelich unpublished data). The 100-ha blowdown–burn combination has a much larger effective disturbance size for sugar maple than a 1000-ha blowdown where maple seedlings are everywhere within the disturbance perimeter.

Does size of severe fires make a difference in the near-boreal forest? Yes and no. For jack pine and the associated black spruce the answer is usually no. Observed post-fire composition of the dominant species is the same – jack pine and black spruce – after an exceptionally wide range of fire sizes: 1000 ha in 1976; 21000 ha in 1910; 91000 ha in 1875; and 180000 ha in 1864 (Heinselman 1973, 1981a,b, Ohmann and Grigal 1981, Frelich and Reich 1995b; Romme *et al.* 1998). For most of the area within these fire perimeters, effective disturbance size was zero, since millions of serotinous seeds rained down on nearly every hectare of the forest floor after each fire. There is one exception, however: part of the 1864 fire was reburned by the 1875 fire (Figure 6.7). This overlap area of several thousand hectares works on the wave-form model for all species, including jack pine, which was eliminated from the area of burn overlap (Heinselman 1973). For other near-boreal species, including balsam fir, red pine, white pine and white cedar, the answer as to whether fire size matters is – to make things complicated – once again yes and no. In theory, size of severe fires matters a lot for all these species, which do not sprout or have canopy-stored serotinous seeds (Romme *et al.* 1998). All are exterminated from areas severely burned, resulting in wave-form succession after fire. However, in the particular case of the landscape of interest – the Saganaga Batholith within the Boundary Waters Canoe Area Wilderness – there are so many refuges from fires, especially lakeshores, that 90% of the area is within 100 m of water. The average fire size is 4000 ha and significant fires are more on the order of 20000 ha to 160000 ha (Heinselman 1973). Thus, fires are huge compared with the interlake distance, and the effective disturbance size for these species is controlled by the placement of lakes, which controls the placement of surviving individuals more than the location of fire perimeters. Succession is wave-form in nature with most of the surviving trees in a belt about 15 m wide along rocky lakeshores (Frelich and Reich 1995a, and Figure 6.13).

To summarize, we may refer to severe disturbances as 'stand-replacing'; however, there is often a legacy of surviving individuals within. Thus, the term stand-replacing applies more to the demise of the former canopy of mature trees than to the species in many cases. Whether there is wave-form succession or in-place succession varies according to complex interactions among disturbance types and tree species.

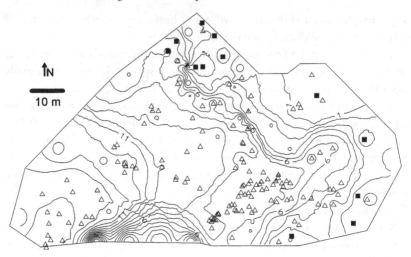

Figure 6.13. A case study in wave-form succession: Explosion Island in Seagull Lake, BWCAW, Minnesota. Solid squares show the locations of large red and white pines that survived a fire 8 years prior to mapping the island. Note that surviving pines are all within 15 m of the lake. Triangles represent paper birch and pin cherry in the island's interior that came in by seed after the fire. Contour interval 1 meter, starting from the waterline. After Frelich and Reich (1995b).

Conifer–hardwood (or evergreen–deciduous) mosaics

Causes of mosaics

Diverse and striking vegetation patterns – mosaics of contrasting species with patches at the neighborhood and stand scales – occur in temperate forests (Table 6.4). We already know that such mosaics can be caused by a complex interaction among disturbances, succession, and physiographic site factors (He and Mladenoff 1999, Figure 6.14). However, there is one more factor shown in Table 6.4 that contributes to the patchy nature of conifer–hardwood mosaics: neighborhood effects. It is not possible to have a complete understanding of forest patch structure without taking this widespread and important force into account. Previous discussions in this book have examined how species biotic properties interact with disturbance to alter successional trajectories, but have not addressed the topic of biotic properties as a cause of patch formation. Biotic properties by which individual trees alter their immediate surroundings are known as neighborhood effects (Frelich *et al.* 1993, Frelich and Reich 1995a,b, 1998, 1999, Ponge *et al.* 1998). The formal definition of neighborhood effects in forests is:

Table 6.4. *Causes of conifer–hardwood mosaics and some examples of each*

Cause of patch differentiation	References
Different soil or ecosystem types support conifers within a hardwood forest	
Pine or hemlock on a pocket of sandy soil	Pastor *et al.* 1984
Black spruce stand on a sphagnum bog	Curtis 1959
White cedar stand on an alkaline wetland	Curtis 1959
Climate where high-elevation cool spots support conifers within a hardwood forest	
Red spruce on hilltops in New England	Leak 1975
Climate where high-elevation warm spots support hardwoods within a conifer forest	
Sugar maple in northern Great Lakes	Figure 6.14
Summer cool lake-effect climate supporting conifers within hardwood forest	
White cedar on bluffs near Great Lakes	Curtis 1959
White spruce in northern Great Lakes	Curtis 1959
Disturbance patches where successional species are deciduous within a conifer or mixed forest	
Aspen after two fires burn jack pine forest within 10 years on poor soil	Heinselman 1973
Aspen after a single fire in boreal forest on good soil	Bergeron *et al.* 1998
Aspen–birch after windthrow–fire combination in hemlock–hardwood forest	Frelich and Reich 1999
Disturbance patches where successional species are conifers within a hardwood forest	
Pine and hemlock within sugar maple forest after surface fire	Frelich and Lorimer 1991a
Pattern of invasion	
Hemlock invades wet-mesic areas within mesic sugar maple forests	Frelich *et al.* 1993
Hemlock patches persist in their Holocene pattern of invasion, even though original cause for pattern may no longer exist	Davis *et al.* 1998
Positive neighborhood effects when two species or species groups growing together both possess these effects	
Hemlock and sugar maple or beech mosaic	Frelich *et al.* 1993, Pacala *et al.* 1996
White cedar–paper birch mosaic on the Minnesota North Shore	Cornett *et al.* 1997

Figure 6.14. A conifer–hardwood mosaic caused by Lake effect climate and topography: the Minnesota north shore of Lake Superior. Aspen and spruce forest (on relatively cool lowlands near Lake Superior, foreground), and sugar maple forest (on relatively warm ridge away from lake in background). Photo: University of Minnesota Agricultural Experiment Station, Dave Hansen.

> Any process mediated by canopy trees that affects the replacement probability by the same or other species at the time of canopy mortality. Neighborhood effects are defined in relation to dominant tree species or groups of species. Positive neighborhood effects (analogous to feedback effects) are processes that promote self-replacement; negative effects are processes that deter self-replacement (unless no other species are available); and neutral effects are processes that neither favor nor disfavor self-replacement.

It is apparent that conifers and hardwoods have contrasting neighborhood effects and other life-history and stand characteristics that often times favor their own reproduction and reduce the other's success underneath their own canopies (Table 6.5). According to the definition both groups have positive neighborhood effects.

Two types of neighborhood effects have been identified (Frelich and Reich 1995a). The first type are overstory–understory effects, which can be positive or negative and operate by influencing the species composition of seedlings and saplings underneath canopy trees, which in turn may translate into influence on the species of the tree(s) that replace a

Table 6.5. *Mechanisms that have been proposed for overstory–understory neighborhood effects that may allow differential success of conifer and hardwood seedlings in some circumstances*

Type of neighborhood effect	References
Nutrient availability	
Nitrogen mineralization in duff and upper soil horizons (low under conifers, high under hardwoods)	Mladenoff 1987, Pastor et al. 1984, 1987, Boettcher and Kalisz 1990, Ferrari 1999
Type of nitrogen compounds available (NO_3 in hardwoods, NH_4 in conifer stands)	Kronzucker et al. 1997, Stark and Hart 1997
Differential availability of calcium and other bases (low in conifers and high in hardwoods)	Finzi et al. 1998
Light characteristics	
Density and duration of shade (higher in conifer stands)	Canham et al. 1994
Duff and coarse woody debris characteristics	
Coarse woody debris seedbed availability (high in conifer stands, low in hardwood stands)	Cornett et al. 1997, 2000, Simard et al. 1998
Duff physical characteristics (thick in conifers, coarse with many layers in hardwoods)	Ahlgren and Ahlgren 1981, Cornett et al. 1997, 2000, Simard et al. 1998
Different biota of forest floor in conifer and hardwood stands	
Mycorrhizal differences	Alvarez et al. 1979

canopy tree when it dies. These include such effects as shading, changing the physical and nutrient make-up of the litter layer, stump sprouting and seed rain (Frelich and Reich 1999). Disturbance-activated neighborhood effects – the second type – operate mainly in forests perpetuated by intense fire where seedlings are mostly killed at the same time as the canopy trees. Serotinous seed rain from dead jack pine and sprouting from underground rootstocks of aspen are examples.

In practice, several of the causes of mosaic formation shown in Table 6.4 usually operate on the same landscape, and it may be difficult to separate the contribution of each to the patch-structure of the landscape. Cases are not always so clear as for black spruce on sphagnum-filled bogs embedded within a sugar maple forest. Clearly, the environment of the bog, in terms of low N and Ca availability and wetness of the soil, is beyond the tolerable limits for sugar maple. For fuzzier cases of species mosaics, one hypothesis is that species sort themselves out into patches over long periods of time, based on minor environmental differences. However, these studies always suffer from circularity because it is not possible to show whether the environment was initially different, or the species created environmental differences over time. Experimental establishment of communities on a uniform environment could address the hypothesis, but is impractical in forests. An alternative hypothesis is that in large relatively flat regions like much of the world's temperate forest, neighborhood effects may be a major force causing patch formation and the other environmental factors work to enhance neighborhood-effect differences. For example, the positive neighborhood effects of sugar maple may be reduced or eliminated on sandy soils, allowing hemlock to dominate those locations. Simulations (Frelich *et al.* 1998b) show that very small differences in the strength of positive neighborhood effects – possibly too small to measure adequately in the field – may lead to switches in dominance from one species to another over a few millennia. In any event, it is necessary to understand the potential of neighborhood effects to create patches before trying to determine the relative contribution of various causes of patch-formation.

Measuring neighborhood effects in the field

The overstory–understory type of neighborhood effects can be measured by examining the relationship between overstory composition and understory seedling or sapling composition in neighborhoods with a wide range of overstory composition. The main method used to date consists of nested sub-plots where understory trees are tallied for a small radius nested within a larger radius (or neighborhood) of overstory trees that are judged to influence the understory. The neighborhood radius should be scaled to each forest: about 2–3 times the average crown radius, or large enough to include an average of 10–12 overstory trees, and the nested radius should be at least one crown radius shorter than the neighborhood radius (Lorimer 1983b, Frelich *et al.* 1993, Frelich and Reich 1995a). After the composition of enough neighborhoods has been measured, the strength of the understory–overstory relationship can be analyzed by regression. Frelich and Reich (1995a) related understory sapling density to overstory basal area proportion for dominant species in three contrasting forest types: hemlock, with positive effects indicated by a slope significantly higher than 0; white pine, with neutral effects indicated by a slope not significantly different from 0; and jack pine, with

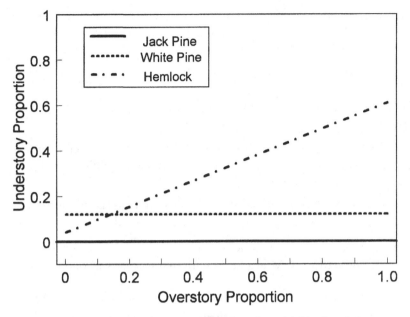

Figure 6.15. Overstory–understory relationships from neighborhoods 9 m or 10 m in radius. The three lines illustrate positive neighborhood effects (hemlock, $n = 48$, from Sylvania Wilderness Area, Michigan), neutral (white pine, $n = 34$, from Hegman Lake, BWCAW, Minnesota), and negative (jack pine, $n = 36$, from Seagull Lake, BWCAW, Minnesota). After Frelich and Reich (1995b).

negative effects in this case indicated by 0 seedlings regardless of the amount of jack pine in the neighborhood (Figure 6.15). The same techniques can be used to detect disturbance-activated neighborhood effects but one must obtain the overstory composition prior to stand-replacing disturbance, and the sapling composition after stand-replacing disturbance in each neighborhood.

Characterizing patches caused by neighborhood effects
Landscapes dominated by trees with positive neighborhood effects have very interesting patch dynamics that in some cases can be independent of disturbances. The MOSAIC simulation was developed to show what types of patch structure could occur when neighborhood effects are the only processes in action within a closed-canopy forest (Frelich *et al.* 1993, Frelich *et al.* 1998b). MOSAIC uses a vector that represents the probability that each species in the forest will replace a dying tree at the center of

a neighborhood. This vector has the dimensions $n \times 1$ ($n =$ number of species), and sums to 1.0 (i.e. there is a 100% chance that a tree of some species will replace a dying tree):

$$SRV = (NRM)(NV) \tag{6.1}$$

where:

SRV is the species replacement vector ($n \times 1$)

NRM is the neighborhood relationship matrix ($n \times n$), which expresses the probability that each species will be replaced by itself if the neighborhood is entirely occupied by that species

NV is the neighborhood vector, ($n \times 1$), that contains the proportion of trees of each species in the surrounding neighborhood

A numerical example will help make clear how this vector is calculated. Suppose that NRM in a forest with two species is as follows:

Matrix 1

	Existing canopy tree	
Replacement trees	Species 1	Species 2
Species 1	0.9	0.1
Species 2	0.1	0.9

And further suppose the neighborhood composition (NV) is:

Matrix 2

Neighborhood composition	
Species 1	0.70
Species 2	0.30

Then SRV is $(0.9)(0.7) + (0.1)(0.3)$ for species 1, and $(0.1)(0.7) + (0.9)(0.3)$ for species 2:

Matrix 3

Probability of replacement by	
Species 1	0.66
Species 2	0.34

In the MOSAIC simulation this process is repeated throughout the forest for the neighborhood surrounding each tree at the time it dies, using realistic tree density, mortality rates, and a specified neighborhood radius. Note that the diagonal of the NRM matrix shows the probability that a tree will replace itself if the neighborhood is entirely occupied by conspecifics (in this case 90%). This is referred to as neighborhood strength. At neighborhood strength of 0.5 the two species would replace each other at random, and at 1.0 no other species would have any chance if the neighborhood was entirely occupied by conspecifics.

When a variety of these simulations was run in a sensitivity analysis, starting with a random mixture of two species, it was shown that neighborhood strength is directly tied to degree of patch formation. The species share much less perimeter and have much more interior patch area as neighborhood strengths increase from 0.5 to 1.0 (Figure 6.16). The characteristics of patches also respond to changes in neighborhood radius (directly related to the number of influential neighbors): small radii result in very compact patches, while allowing a larger number of nearby individuals to influence the chance of replacement results in larger patches with diffuse edges that have many small satellite patches near their edge (Figure 6.17).

Sylvania Case study

Discovering the combination of causes for patches in a conifer–hardwood mosaic is complicated and a lot like detective work. Many different forms of analysis and lines of evidence are needed and these will be different in every case. Now, however, we have all the tools necessary to analyze the Sylvania landscape: characteristics of fires and windstorms (Chapter 2); methods for reconstructing stand history (Chapter 3); knowledge of succession in hemlock–hardwood forests (Chapter 4); and knowledge of causes of patch structure and neighborhood effects (earlier this chapter). Therefore, perhaps this whole process is best illustrated by going through a case study.

Sylvania Wilderness Area (see Chapter 2 'The principle never-logged forest remnants', for description) has a striking mosaic of conifers (mostly hemlock, but also some black spruce in lowlands) and hardwoods, mostly sugar maple, yellow birch and basswood (Figure 6.18, Frelich et al. 1993). Preliminary investigation revealed that at least two causes of patch formation were taking place. Surface fires are common along some lakeshores which have a convex slope as they approach the water's edge. These steep

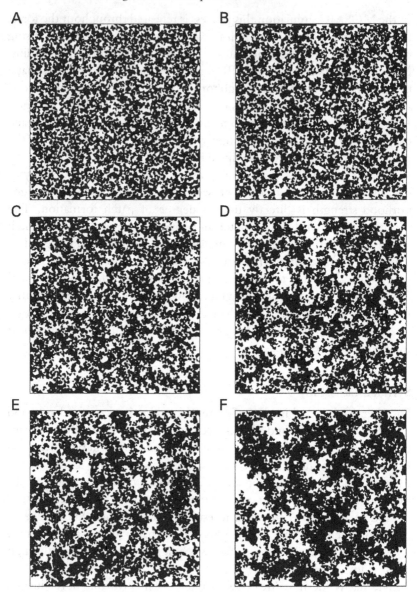

Figure 6.16.   Simulated effects of neighborhood strength on patch structure. Neighborhood effect strengths are 0.5, 0.8, 0.9, 0.95, 0.975, and 1.0 in parts A–F, respectively. After Frelich *et al.* (1998b).

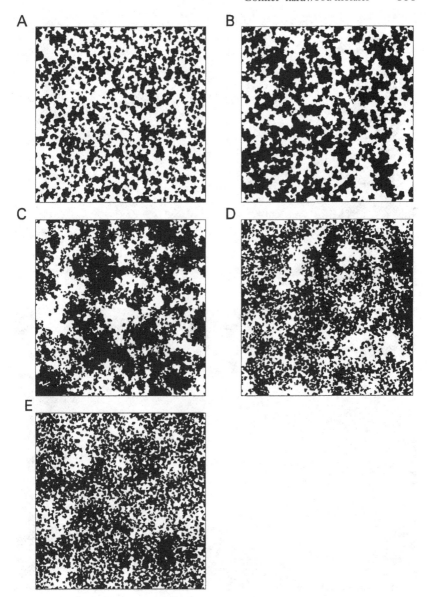

Figure 6.17. Simulated effects of neighborhood size on patch structure. Neighborhood radii of 2.0 m, 5.0 m, 10 m, 15 m, and 20 m, in parts A–E, respectively. After Frelich *et al.* (1998b).

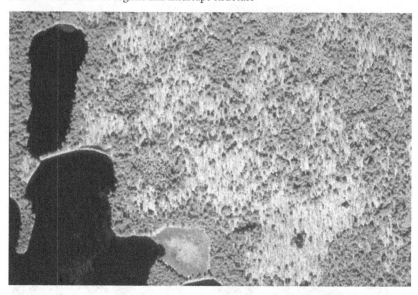

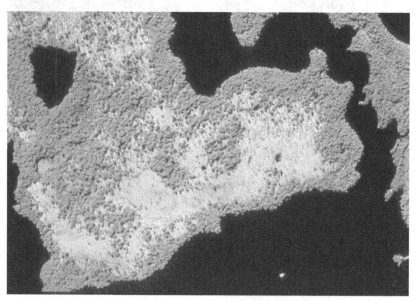

Figure 6.18. Three mechanisms of patch formation in hemlock–sugar maple forests in Sylvania Wilderness Area, Upper Michigan. Hemlock and other conifers appear dark, while sugar maple and other hardwoods appear light in these leaf-off infra-red aerial photographs. Upper photo, patch structure in the middle is formed by a combination of pattern of invasion and neighborhood effects, and hemlock patches in the lower middle are due to the presence of a nearby bog. Lower photo, the strip of hemlock forest along the lake edge was formed after a fire during the 1930s.

slopes receive wind from the lake and are very well drained, leading to a high chance of fires. The fires are naturally extinguished as they work their way into the more humid forest interior. The fire along one lake was indicated by abundant charcoal and fire scars on surviving mature trees, and was characterized by a mixed stand of young hemlock with some paper birch (Frelich unpublished data). This young fire-caused stand has enough hemlock in it that it shows up as a conspicuous dark band along the water's edge on Figure 6.18, lower photo. A second lake-side band of conifers exists because of the wet soil, in areas where the approach to the lake is concave in shape (Figure 6.18, upper photo).

A third area with hemlock and sugar maple patches of many sizes, in the middle of Figure 6.18, upper photo, was initially a mystery. Soil analyses found no consistent differences in parent material or nutrient status in the two patch types (Frelich et al. 1993). Reconstruction of stand disturbance history for a 5-ha plot stretching across the boundary of two of the major patches showed that the hemlock and sugar maple patches were both old multi-aged stands, with the same disturbance cohorts present in each stand, and no indication of differences in disturbance severity in either stand for the last 200 years (Frelich and Graumlich 1994). Analysis of overstory–understory relationships, however, revealed very strong positive neighborhood effects for both dominant species. Output from the MOSAIC simulation with 1.0 neighborhood strength and 10 m neighborhood radius exhibits the same spatial characteristics as the actual forest (cf. Figure 6.18, upper photo and Figure 6.16F). Simply getting a result from a simulation that looks like the real landscape does not prove cause and effect. In this case, however, direct proof is not possible. One would have to observe the actual success of seedlings at the time the patches formed to get at the cause. Therefore, we arrived at the most likely cause of patch formation by a process of elimination. All known potential causes of patch formation other than neighborhood effects were systematically eliminated. At the same time we demonstrated that strong neighborhood effects exist in the field and we used the MOSAIC simulation to show that neighborhood effects are a feasible cause of patch formation.

Looking at a longer temporal scale, however, there is still a problem with this chain of reasoning. The patches could have been caused by some other factor long ago and simply be maintained or strengthened by neighborhood effects at this point in time. Fortunately, there are many small hollows in Sylvania that have a well-preserved sedimentary sequence containing fossil pollen and sometime macrofossils (Davis et al. 1994, 1998).

These hollows reveal that the current mosaic of hemlock and sugar maple-dominated stands was established shortly after hemlock invaded Sylvania 3100 ybp (Davis *et al.* 1994). At the time hemlock invaded, the forest was already a hardwood–conifer mosaic, but with different species. White pine patches alternated with patches dominated by oak (possibly red oak) with some sugar and red maple. Hemlock apparently found better conditions for regeneration in the pine stands and preferentially invaded those patches (Figure 6.19). Once hemlock was present the frequency of fire dropped, not only in the hemlock patches, but on the entire landscape. In the absence of fire, sugar maple was able to replace the red maple and oak, and white pine began to gradually disappear from one mixed hemlock–white pine stand after another – a process which still continues today. At this point we don't know why the original pine–oak mosaic existed. But the landscape 3000 years later obviously has some memory of those patches. We can summarize this case study by saying that we know that there are patches of hemlock caused by fire and by areas of wet-mesic soil, and by a combination of pattern of invasion and neighborhood effects (Figure 6.19).

Summary

Disturbance regimes are one of the major forces that structure the mosaic of forest communities across the landscape. Therefore landscape characteristics are sensitive to changes in disturbance regimes. Disturbance regimes dominated by wind are generally dominated by late-successional species and wind creates a very complex web of stands in many stages of development. For example, the hemlock–hardwood forests of the Lake States have eight different structural stages with complex, individualistic routes of development for each stand. The proportion of landscape in young even-aged stands and steady-state stands is very sensitive to windstorm frequency. The size distribution of trees across the landscape is less sensitive to changes in windstorm frequency.

Fire is different from wind in that it more easily regulates the distribution of species across the landscape. Fire is important for maintaining diversity of tree species even in landscapes where fire is rare. For example, red oak and white pine depend on the occurrence of rare fires to maintain a presence within hemlock–maple-dominated forests. In regions with frequent fire, such as the near-boreal forest of northern Minnesota, fire also regulates species composition but in a different way. Fires occur at random with respect to stand age, so that some stands burn twice in a

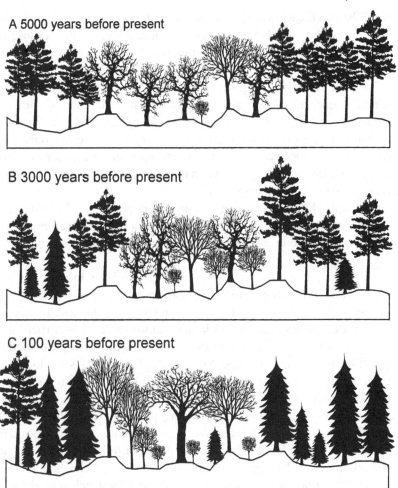

A 5000 years before present

B 3000 years before present

C 100 years before present

Figure 6.19. Idealized development of the hemlock–hardwood patch structure on the Sylvania landscape over time. A, a mosaic of white pine and oak forest with some sugar maple existed about 5000 years before present. B, about 3000 years before present hemlock began to invade the pine patches, shutting off the flow of fire across the landscape and allowing sugar maple to begin replacing oak in the hardwood patches. C, at the time of European settlement hemlock had replaced nearly all of the white pine, and sugar maple had replaced nearly all of the oak.

short time and others are skipped for two or more rotation periods. As a consequence, most stands burn between the ages of 20 and 120 years, leading to continued dominance by jack pine. Other stands do not burn for more than 150 years, allowing succession from jack pine to black spruce, balsam fir, white cedar and paper birch on small patches across the landscape. Other stands burn when less than 20 years old, leading to replacement of jack pine by aspen. Thus, fire regulates a landscape mosaic of jack pine, spruce–fir–cedar–birch, and aspen stands.

Complex disturbance regimes with wind, fire and herbivory occurring all at once regulate forest composition through a set of complex interactions among disturbance types. For example, white pine in northern Minnesota must get large enough to survive the next surface fire, otherwise continued recruitment of white pine into the canopy cannot be insured. Deer browsing may retard the growth of seedlings so that they never get large enough to survive the next fire. If a windstorm then removes old white pines from the canopy the species will be extirpated from the stand and only reinvasion from an adjacent stand can restore it. If this scenario is common, then few stands across the landscape will have white pine.

Succession is a spatial process. Each species may have 'in-place succession' or 'wave-form succession' after a disturbance type, depending on whether individuals or propagules are present after the disturbance or whether the species must reinvade after disturbance. When there is a species–disturbance combination that works on the wave-form model, then disturbance size matters. Otherwise, disturbance size does not matter because the species survives at numerous points within the disturbance perimeter. Species that cannot sprout after fire, have no serotinous seeds, soil-stored seedbank, or surviving seedlings or adults within a given disturbance will all create wave-form succession as they respond to disturbance. Blowdown size does not control composition in hemlock–hardwood forests because seedlings survive the blowdowns. Fire size does not control composition in near-boreal jack pine because the species survives within the fire perimeter as serotinous seeds. Effective disturbance size is zero in these two cases.

Neighborhood effects between tree species also interact with disturbance to structure the landscape patch mosaic. For example, hemlock and sugar maple form patches 2–20 ha in size by excluding each other's regeneration from beneath their canopies. Disturbances such as fire and changes in soil type can magnify neighborhood patch-forming processes.

7 · Disturbance in fragmented landscapes

Many formerly forested landscapes around the world are now fragmented. In this chapter I show how fragmentation alters the disturbance processes discussed in previous chapters. The emphasis is on 'external forest fragmentation', which applies to situations where most of the landscape has been converted to non-forest. Small islands of forest called 'woodlots' exist in a sea of agricultural fields or suburban developments (Figure 7.1). Fragmentation may also exist when the 'fragment' is not completely surrounded by human landscape elements. Disturbance processes operate differently in these fragmented environments. Curtis

Figure 7.1. Landscape dominated by agriculture with small fragmented woodlots in southern Minnesota. Photo: University of Minnesota Agricultural Experiment Station, Dave Hansen.

(1956) was among the first to recognize that a species depending on mature forests for existence could be lost after a major disturbance on a fragmented landscape, because there would be no source of propagules from the surrounding lands. Fragments can be steadily lost as they are disturbed over time, almost as if they are being 'mined' by disturbance until none is left. This loss is not solely due to the immediate disturbance effect, but also because fragmentation limits the recovery of native species via complex mechanisms. Fragmentation by itself can change the frequency of fires without human fire suppression due to a fragmentation-dilution effect on disturbance frequency.

Forest fragmentation and herbivory

Development of alternative communities caused by herbivory

Several authors have recently suggested that stability of plant–herbivore equilibria depends on both plant and herbivore density so that two alternative stable states can exist for a single herbivore density (Noy Meir 1975, May 1977, Dublin *et al.* 1990, Schmitz and Sinclair 1997). This phenomenon occurs in fragmented maple–oak forests inhabited by white-tailed deer in the 'Big Woods' of southern Minnesota (Augustine *et al.* 1998). Alternate forest types with and without lush forest-floor plant communities have developed. In some stands the understory is so lush that even the highest deer density in this region cannot control the plants. However, in other stands, where the lushness of the understory is below a certain threshold, a moderate deer density can drive the understory biomass to nearly zero. A study of the dominant understory plant wood nettle (*Laportea canadensis*) was carried out in these forests (Figure 7.2). The models of Noy-Meir (1975) and May (1977) predict a non-linear response of vegetation to increasing browsing pressure, and this non-linear response probably explains the separation of forests into two types, rather than a continuum. The basic model for changes in plant abundance (Noy Meir 1975) is described by

$$\frac{dV}{dt} = G(V) - c(V)H \tag{7.1}$$

where:

V is plant abundance

$G(V) = r\ V(1 - V/K)$ is the logistic growth function where r is the intrinsic rate of increase, and K is the carrying capacity

Figure 7.2. The wood nettle (*Laportea canadensis*) forms a dense understory in many Minnesota sugar maple–dominated forests. Photo: University of Minnesota Agricultural Experiment Station, Dave Hansen.

$c(V)$ is the herbivore functional response

H is a constant herbivore density

The predictions depend on the herbivore's functional response to plant availability. This functional response of wood nettle to deer herbivory was found to be type II in the Minnesota forest (Augustine *et al.* 1998), a monotonically descending concave form:

$$c(V) = cV / (1 + cHV) \qquad (7.2)$$

A curve of this form dictates that per plant impact of browsing continues to increase as plants get less abundant, indicating that deer encounter no difficulty in finding the plants as they become rare. Two points of interest occur along this curve: one at which both herbivores and high plant

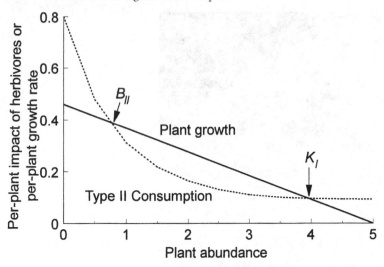

Figure 7.3. Type II consumption function and predicted equilibria under conditions of constant herbivore density. A high plant-density equilibrium (K_I) where herbivores limit plant density is possible. This function may also have an unstable equilibrium (B_{II}) such that initial plant densities below this level lead to plant extirpation (i.e. consumption is greater than plant growth). After Augustine *et al.* (1998).

density can co-exist, and another at an unstable equilibrium such that initial plant density below that level leads to extirpation (Figure 7.3). Note that in fragmented forests deer get most of their food from nearby farm crops and return to the forest for cover where they continue incidental browsing of tree seedlings (during winter) and grazing of herbaceous understory plants (during the summer). Herbivore density is controlled by forces outside the forest and can continue to be high even while forest plants are being extirpated. Hence the term H, a constant herbivore density, in the equation above works for the fragmented forest situation.

One can verify experimentally whether this scheme describes the actual process of deer–plant interactions in the forest. If so, one would expect the following deer exclosure experimental results over time:

1. In stands with low deer density and high plant density, deer should not be regulating plant abundance or reproductive success. Therefore, there should be no significant difference in plant abundance inside and outside exclosures.
2. In stands with high deer density and low plant density, deer are assumed to regulate plant density. Therefore, plants should be much more successful inside exclosures than outside.

3. In stands with high deer density and high plant density, deer are assumed not to be regulating plant density. Therefore, there should be limited differences between plant success inside and outside of the exclosure.
4. In stands where deer and plant density are both low, deer are assumed not to be regulating plants, and there should be no difference in plant success inside and outside the exclosures.

The actual deer exclosure data for wood nettle in southern Minnesota oak–maple forests was consistent with these experiments, although the fourth experiment could not be done because conditions of low deer and low plant abundance could not be found (Augustine et al. 1998). Using a discrete version of equation 7.1 above, with fitted parameters from the field data, and a range of deer densities that occur in the region, a series of simulations was done. The results show that in those stands where deer density is low, wood nettle abundance is likely to increase over time regardless of the initial plant abundance. For moderate-to-high deer densities, however, a breakpoint occurs between 0.1 and 0.6 wood nettle stems per m^2 (Figure 7.4). Above those densities, wood nettle populations are predicted to rise to carrying capacity, while below those densities wood nettle will eventually be driven to extinction.

If these dynamics are common in fragmented forests, then there are four lessons we can learn: (1) herbivore impacts on plant species (and tree seedlings) will be most severe when plants are rare; (2) management of the herbivore population can create a condition where target plant species and herbivores are both abundant; (3) small changes in herbivore or plant populations in cases where they are both abundant can cause a sudden crash in the plant population; and (4) in areas where deer are abundant, restoration efforts may require either establishment of massive numbers of plants (note the scale on Figure 7.4, 1 plant/m^2 is 10 000 per hectare), or fencing out the deer until plant density recovers to a level above the estimated threshold.

The habitat-island effect and deer: hemlock in Upper Michigan

The presettlement landscape with hemlock spreading over thousands of km^2 has been changed so that there are now small islands of hemlock within a sea of second growth aspen and paper birch forest. However, because the area has cold winters and the snowpack is quite deep, deer use the remaining isolated hemlock stands for protection from wind and to take advantage of the lower snow depth under evergreens, where the

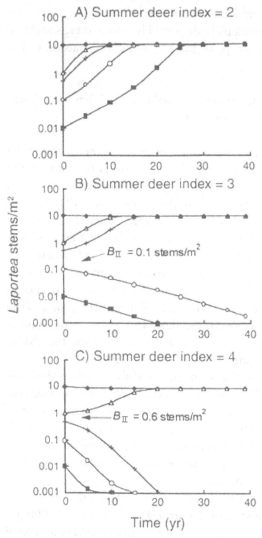

Figure 7.4. Simulated changes of wood nettle populations over time, given different deer densities and initial densities of wood nettle. Deer density is assumed to remain constant over time. From Augustine *et al.* (1998).

canopy intercepts much of the snow. Thus, the concentration of deer in these hemlock stands in winter is very high. Combine this with the facts that hemlock seedlings are the only green browse available during the winter, that deer prefer hemlock as food over other tree species in any case, and that hemlock is not capable of resprouting after browsing, and this situation leads to destruction of the seedling layer on the forest floor. A forest type which would normally be stable for centuries or millennia is becoming unstable because deer are preventing reproduction of hemlock for the long term. There has been no significant establishment of hemlock seedlings since the 1920s in the area and at this point sugar maple is replacing hemlock (Frelich and Lorimer 1985).

Loss of old-growth remnants to wind: risk analysis

Many governmental units in the Lake States attempt to maintain a system of 'natural areas' that represent the various natural communities of the region. This is true for old-growth forest remnants, many of which are small and isolated. As time goes on, these remnants are slowly damaged by deer as explained above, and they are also lost to stand-leveling windthrow.

A spatial simulation was developed, using realistic forest blowdown sizes (Figure 2.2), a variety of forest remnant sizes, and a variety of rotation periods that encompass the variability of severe windstorm frequency across the Lake States (Frelich et al. 1998a). The simulation illuminates the interaction between forest fragment size and chance of being hit by disturbance. The smaller a forest remnant is the less likely that it will be hit by high winds. If a small forest remnant is hit, however, the canopy is likely to be totally destroyed. Large remnants are very likely to be hit by high winds, but are less likely to have the canopy totally leveled. Even when one of the largest downbursts (about 6000 ha) hits a 5000 ha forest remnant, the likelihood that the alignment of the storm will exactly match the alignment of the forest remnant is remote. However, more than one storm could impact the area.

The simulation results show that the depletion of small forest remnants (1 ha) by high winds over time is essentially the same as the proportion of the landscape disturbed over time in a continuously forested landscape (Figure 7.5). For example, with a rotation period for stand-leveling winds of 1000 years, about 10% of small remnants would be lost each century. One-hectare forest remnants are essentially the same as a point when compared with the whole landscape. Remnants this small are not likely to recover after disturbance if they are surrounded by agricultural fields, because a number of alien species are likely to invade immediately after disturbance. On the other hand, remnants over 5000 ha in size are virtually immune to loss from high winds (Figure 7.5). There is almost no chance that all or most of such a remnant will be leveled over a century. The prospect for recovery of those portions that do blow down is good, because there will always be some remaining intact forest adjacent to the disturbed area (Peterson and Carson 1996).

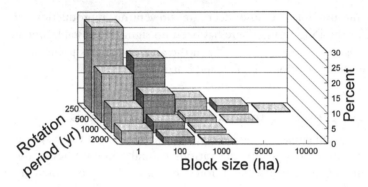

Figure 7.5. Probability of losing primary forest remnants to windthrow, depending on rotation period and size of forest-fragment size. Percentage of blocks experiencing more than 80% blowdown over one century (assuming the historical size distribution of blowdowns from Figure 2.2) is shown.

The disturbance–dilution effect of fragmentation

Fragmentation dilutes the occurrence of any disturbance type that requires intact forest for its spread. In other words this effect would apply to fire but not wind, which blows whether there is surrounding forest or not. The magnitude of the dilution effect depends on the relative size of disturbances and forest fragments. A simplified example should serve to illustrate the effect. Imagine a large (1 000 000 ha) intact forest landscape where all fires are oval in shape and 10 000 ha in size, the rotation period is 100 years, and the hazard function is uniform with respect to stand age and location. On average, one of the 10 000 ha fires (covering the required 1% of the landscape) will occur each year. Further imagine a 1000 ha tract of forest in the middle of this forest (Figure 7.6). There is a 1% chance every year that fire will burn part or all of the 1000 ha tract. However, there is only a 0.1% (or 1/1000) chance that the fire will actually start within the 1000 ha tract. Now imagine that the 1000 ha tract is a remnant surrounded by farmland. It will only burn now if a fire starts within the 1000 ha. Assuming the chance of ignition is still equal on an area basis (this may be the case if fires are caused by lightning), then a fire will ignite within the 1000 ha tract only once every 1000 years. The rotation period would then be 1000 years or more, since not all fires starting within the tract would burn the entire area (Figure 7.6).

This dilution effect of fragmentation is presumably why islands in lakes burn less often than mainlands. There is a trade-off in the case of islands whereby they may be more likely to be struck by lightning than mainland

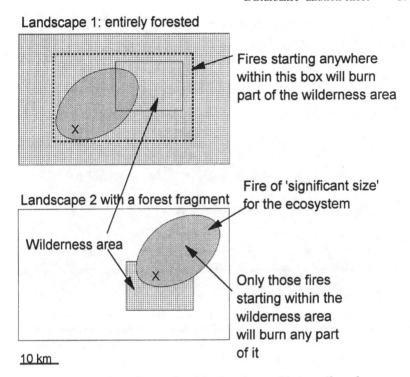

Landscape 1: entirely forested

Fires starting anywhere within this box will burn part of the wilderness area

Fire of 'significant size' for the ecosystem

Landscape 2 with a forest fragment

Wilderness area

Only those fires starting within the wilderness area will burn any part of it

10 km

Figure 7.6. Hypothetical example of the disturbance-dilution effect of fragmentation. Shaded areas indicate forest.

areas, because trees on them are the local high point. However, if the island is very small, the disturbance-dilution effect of fragmentation may quite easily override the countervailing effect of increased chance of lightning strikes and result in the decreased probability of burning noted by Heinselman (1973) for small islands in the Boundary Waters Canoe Area Wilderness (BWCAW). As a result of lower frequency of burning, the islands also had a different species composition than those areas that burn more often (Figure 7.7).

One-sided fragmentation can also result in substantial reductions of fire frequency, especially when fires are very large. This scenario has happened in the BWCAW, where inspection of Heinselman's area burn maps (Heinselman 1996) reveals that almost all major fires within the wilderness area during the presettlement era from 1600 to 1900 have their long axes arranged in a south-to-north, or southwest-to-northeast direction. Because the wilderness area is long and thin in an east-to-west direction these fires used to burn clear through the wilderness and into

Figure 7.7. Red and white pine on an island in the Boundary Waters Canoe Area Wilderness, a region where most mainland sites historically burned too frequently to allow these species to dominate. Photo: University of Minnesota Agricultural Experiment Station, Don Breneman.

Canada (Figure 7.8). Now that the area to the south of the wilderness has been converted to less flammable forest types (aspen instead of conifers) and other land uses, such as highways, towns and resorts, these fires that start outside the wilderness and later burn into it no longer occur. Because typical significant fires were 30–60 km in length (Figure 7.8), the one-sided fragmentation effect could extend that far into the near-boreal forest. It is the opinion of some that this is the most important contributor to the lengthening of fire rotation periods in the wilderness from 100 years to 1000 years over the last century. This situation may be partially reversed if the forest outside the wilderness succeeds back to conifers, which now depends on the frequency of logging.

Summary

Small fragments of forest can lose their native species over time due to disturbance of some sort followed by invasion of non-native species. Disturbance size makes more of a difference for fragmented forests than for large contiguous forested landscapes because of the relatively high

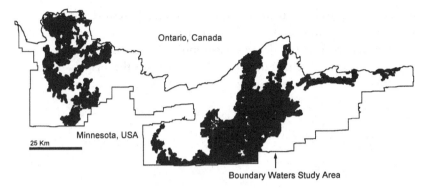

Figure 7.8. Map of the Boundary Waters Canoe Area Wilderness, showing major burned areas during the 1800s (after Heinselman 1973). Note that large fires often started to the south of the wilderness area and burned as much as 40–60 km into it.

availability of propagules of other species from outside the stand after disturbance. Forest fragments are also likely to suffer browsing of the native plant species by deer, although other small mammals such as rabbits and rodents can also contribute to loss of plants. Thus, stand-leveling winds in small fragments can mean loss of stability that hemlock–hardwood forest would usually have after windthrow, and a switch of species dominance that would not occur in an unfragmented landscape. Large fragments of forest (>5000 ha) are not likely to blow down all at once because the orientation of storms is not likely to match the orientation of the fragment. Thus, recovery of native forest after blowdown is more likely in large fragments due to the continued existence of seed sources adjacent to the blowdown.

On landscapes where disturbances are very large, such as the near-boreal forest of northern Minnesota, fragmentation can occur at huge scales. The 400 000 ha BWCAW is fragmented along one side. Because most fires that formerly burned significant areas within the BWCAW came from that side, fire frequency within the BWCAW has been reduced by the disturbance-dilution effect of fragmentation.

High deer populations can be supported by agricultural crops surrounding forests and hunting policies of various resource managers. These deer can prevent regeneration of native forest species to the point of creating a barren forest. It is possible to have a fairly high deer population and a lush forest environment at the same time within a forest fragment, but a high density of plants is required, so that annual growth of plants is more than annual grazing by deer. Restoration efforts that do not take this into account are doomed to failure. To restore native species such as wood nettle in an environment of high deer populations one must either saturate the deer's consumption by planting at least 10000 wood nettles per ha, or fence out the deer until the plants attain that density.

8 · Forest stability over time and space

This concluding chapter is a synthesis of everything said earlier in the book. Here I examine stability of age structure and species composition over time and the different types of dynamics that forests may exhibit as a result of their level of stability. Some of the most important linkages among neighborhood, stand and landscape spatial scales will be made here. The real reason we are interested in the material from all of the previous chapters is to generalize about stability of forests. Under what conditions will they change or stay the same? We need to answer those questions now because we are purposely changing the disturbance regime in forests from one dominated by natural disturbance to one dominated by harvesting. Global climate change will also change the disturbance regime, even in forests reserved from logging, in ways that are difficult to understand.

Sometimes investigators have found that their study site was just big enough for a certain disturbance process to operate in stable fashion, according to the study results (e.g. Lorimer 1980, Lorimer and Frelich 1984, Frelich and Lorimer 1991a; Frelich and Graumlich 1994, Frelich and Reich 1995b). This is because there is a continuum of disturbance processes at overlapping temporal and spatial scales and researchers mostly detect the ones operating at the scale of their study area. Some studies examine a large-scale process in a small study area and conclude that forests are unstable. No one is really right or wrong in this stability debate because like the direction of succession in the near-boreal forest (see Chapter 4), it all depends on how one looks at the situation.

One must never forget that there is always a bigger, more rare, or more severe disturbance type than the ones under study, even if these consist of glaciations that peak every 100 000 years, or comets that simultaneously wipe out all forest ecosystems on earth every 65 000 000 years. Therefore, it is necessary to define the relevant scales at which forest stability could exist. Temperate and cold-temperate forests of the world only exist in their current configuration during interglacials. Because tree species take up about 50% of available time during an interglacial just to migrate into a reasonable approximation of their potential range (although finer-scale adjustments continue due to smaller climate fluctuations), the maximum periods of time available for these forests to develop is on the order of 5000 years. Paleoecological analyses indicate that, at least in this interglacial, periods where the climate is stable enough to perpetuate one forest type with all of its dominant species in place (having completed the rapid phase of their migration), could last 1000–5000 years (Davis 1981, Webb 1987). Therefore, stability at that time scale and all smaller scales can occur. This could include various equilibria at landscape (~1000–1 000 000 ha), stand (~1–10 ha), and neighborhood (~0.01–0.1 ha) scales.

Because stability in age structure and composition behave differently, they are taken up separately in the following sections and then integrated back together at the end of the chapter. For example, one stand of jack pine within the near-boreal forest could achieve an equilibrium composition for centuries, even though age structure would never reach an equilibrium due to repeated burning. Stability in biomass has been dealt with extensively elsewhere (Bormann and Likens 1979, Shugart 1984) and is not covered here.

Stability of age structure

The quasi-equilibrium concept

Stability of age structure at any spatial scale – the minimum dynamic area – can be assessed via the quasi-equilibrium concept (Shugart 1984). The simplest definition of a quasi-equilibrium is that the age-class distribution of a study area of interest remains stable over time. Such a quasi-equilibrium has two subjectively defined requirements: (1) disturbance-caused patches must be small relative to the total area, and (2) disturbances must occur at a relatively constant rate over time. If both of these conditions are met, then the distribution of stand ages across the landscape or the distribution of tree ages within a stand, as the case may be, will meet the earlier discussed criteria for stable age distributions: a flat or monotonically descending shape without major peaks or gaps. The smoothness of the age distribution is the important feature. If the age distribution is plotted as 10-year age classes, the residuals from a smooth-line fit through the curve for each decade should be small. If regression of histogram bars against a fitted curve is used, remember that these are not independent sample points that one would have in a true relationship between two variables. The normal criteria for r^2 to judge the fit do not apply. Also note that fitting the negative exponential via a semi-log procedure makes the fit appear better than it is. Peaks in the age distribution that represent 20–25% of the landscape can appear very minor on a semi-log scale. Therefore, the age-class distribution should be plotted on an arithmetic scale, and one must simply come up with some subjective criterion for the largest residual allowed that still is considered to represent quasi-equilibrium conditions. One criterion already used is that no single disturbance should occupy more than 20% of the total area over a 250-year period (Frelich and Lorimer 1991b). If the observations of stand age or tree age are independent, then the Kolmogorov–Smirnov test statistic can be used to see if the observed age distribution deviates significantly from some theoretical age distribution (Frelich and Lorimer 1991a). This test does not, however, alleviate the need to develop a subjective criterion for what constitutes an equilibrium.

One must also be careful to distinguish cases in which fluctuations in size of age classes over time are due to changing disturbance frequency versus large disturbances. Examination of the landscape spatial patterns are necessary to accomplish this. For example, the age distribution for 46 pooled plots in the Porcupine Mountains shows decades with low and high disturbance rates (Figure 8.1). At first, it was hypothesized that disturbance frequency had changed in response to the ending of the Little Ice Age in the late 1800s, accounting for the peak in disturbance at that point. However, further investigation showed that high-disturbance

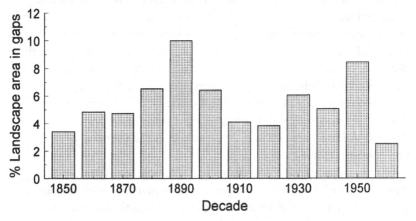

Figure 8.1.   Disturbance frequency from 1940 to 1969 from 46 plots in the Porcupine Mountains, Upper Michigan. After Frelich and Lorimer (1991a).

decades were due to the chance occurrence of a large downburst within the study area in some decades. Downbursts that level about 1000 ha of forest have a frequency of approximately once every 50 years for a landscape the size of the Porcupine Mountains (14500 ha). Landscapes 10× and 100× the size of the Porcupine Mountains would have two and 20 such blowdown(s) every decade, respectively. Thus, these observed fluctuations are caused by the relationships among size of disturbance, size of landscape, and frequency of occurrence.

Age distributions are not always available, and other ways of assessing the stability of the landscape may be necessary. Often, one has knowledge of the overall characteristics of disturbance. Shugart (1984) suggests that a landscape area be ≥50× the average patch size as a reasonable threshold for the above-mentioned size criteria. One may also use several thresholds, such as 5× for low stability of the landscape, 25× for medium stability, and 50× for high stability. However, all of these criteria will not work well unless the size-distribution of disturbance patches is normally distributed. For many natural disturbances, especially fire at the landscape level, the few largest disturbances account for a majority of disturbed area. Therefore, Johnson (1992) suggested an alternative criterion that the study area be 2× the maximum disturbance size. However, this too has a problem, since there is really no maximum disturbance size (remember there is always a bigger, rarer disturbance that could happen). One solution is to take the 'significant disturbance size' for some time period, modeled after the mariners' 'significant wave height' during a certain storm (defined as the average height of the largest one-third of all waves). This statistic tells ship captains what they must really look out for while sailing. To adapt this to disturbance, one may divide disturbances into significant and insignificant ones on the basis of area disturbed. A variety of criteria could be developed, such as 'the average size of those disturbances that account for 90% of all area burned'. In the boreal forest, this would probably take only the top 3–5% of all fires. Since these big fires account for most of the patch area, one would still have to take 50× this size for a quasi-equilibrium landscape, since other disturbances were deemed insignificant and are not included in the estimate of mean significant disturbance size.

On complex landscapes comprising several ecosystem types with different disturbance regimes, natural fragmentation can sometimes lead to more stability for a given ecosystem type, when there are many occurrences over a widespread area. In this case, all or most occurrences are not likely to be destroyed by disturbance all at once, even if they are small

Table 8.1. *Proposed criteria for age-structure stability*

Number of landscape units	Level of stability for landscape age structure		
	Low	Medium	High
1	5×	25×	50×
5	1×	5×	10×
10	0.5×	2×	5×
25 or more	0.2×	1×	2×

Note:
The value in each cell shows how many multiples of significant disturbance size each landscape unit must be for the level of stability indicated in column headings.

compared with the disturbance size. In effect, their fragmentation makes them independent with respect to large disturbances. This is another way of viewing the quasi-equilibrium concept: any configuration that leads to a stable distribution of stand ages is acceptable (Table 8.1). However, whether it is an ecosystem with one large patch so big that disturbances can never disrupt the whole at once or whether there are many independent occurrences, one must always watch for fluctuations in number of disturbances over time. Periods of high disturbance frequency can lead to synchronization of the landscape even when disturbances are small or forest patches are independent.

Stability of age structure and spatial scale

A synthesis of studies reveals that quasi-equilibrium type stability, when it occurs at all, can occur for short periods for neighborhoods (a few decades), intermediate periods for stands (several decades to a few centuries), and long periods for landscapes (centuries to a millennium). Only in forests dominated by late-successional forests does stability of age structure extend through all three scales.

Some very small parcels of forest can be surprisingly stable. Stands 0.5 ha in size in the hemlock–hardwood forest of Upper Michigan sometimes attain the theoretical 'steady state' conditions as indicated by the size structure (Lorimer and Frelich 1984). A 0.5-ha steady-state stand could easily meet the patch-size requirement for a quasi-equilibrium if most disturbances are single tree gaps. The mean unbiased canopy-gap size from Runkle (1982) is about 44 m^2, or 1/114 of 0.5 ha. Also, trees only 5–10 m apart may have independent decades of recruitment, and hence are typically part of different patches (Frelich and Graumlich 1994). Thus, 0.5 ha is large enough to have >50 patches the size of significant disturbances. The STORM simulation indicates that about 3.6% of all stands are in this steady-state condition at any given point in time and these stands stay in that condition for 72 years, on average, under the natural disturbance regime (Frelich and Lorimer 1991b). Stands that are maintained in a multi-aged condition, with criteria for stability slightly relaxed, are the most common on the landscape (Hough and Forbes 1943, Leak 1975, Lorimer 1980, Frelich and Lorimer 1991a).

At the other end of the landscape spatial scale, the 14 500 ha primary forest remnant in the Porcupine Mountains in Upper Michigan may exist in a state of quasi-equilibrium with heavy disturbances caused by thunderstorm downbursts that have a rotation period of about 2000 years, and blowdown as much as 1800 ha, or 12% of the landscape at a time. Such blowdowns, however, only occur once every several decades somewhere within the Porcupine Mountains. Disturbances large enough to disrupt the quasi-equilibrium of the entire Porcupine Mountains do occur, with an estimated rotation period of 4000 to 8000 years (Frelich and Lorimer 1991a).

The average rate of gap formation in the hemlock–hardwood forests also is consistent with the hypothesis of great stability across scales from stands to the landscape. Rates of disturbance estimated for a 5-ha study area in Sylvania Wilderness of 5.4% forest canopy area per decade and the corresponding canopy residence time of 186 years are similar to estimates for much larger areas (Frelich and Graumlich 1994). Three large remnants of hemlock–hardwood forest totaling 23 000 ha analyzed by Frelich and Lorimer (1991a) via two different methods produced average canopy residence time estimates of 175 and 155 years. Runkle (1982) analyzed widespread study areas in Ohio, Pennsylvania and New York that have a forest type similar to the Lake States hemlock–hardwood forest and found annual canopy-gap birth rates at 0.60 % of total land area, corresponding to a canopy residence time of 167 years. Thus, there is considerable evidence that canopy residence times range from 150 to 200 years in natural hemlock–hardwood forests. The similarity in canopy turnover and residence time in study areas from 5 ha to more than 1000 times that size indicates great stability in disturbance rate over a large range of spatial scales.

The previous discussion has shown that periods of stable age structures always have limits. Wilson and Agnew (1992) suggest that the punctuated equilibrium concept could be extended to spatial and temporal processes in ecology. If so, then forest landscapes exist in a state of 'punctuated

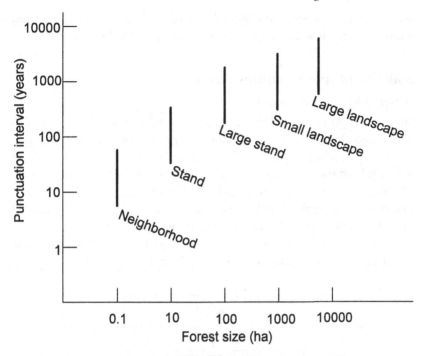

Figure 8.2. Frequency of punctuation events that disrupt quasi-equilibrium age structure for a variety of forest sizes in Upper Michigan's hemlock–hardwood forest. Based on the author's synthesis of data in Frelich and Lorimer (1991a,b), Frelich *et al.* (1993), and Frelich and Graumlich (1994).

quasi-equilibrium', whereby disturbances with small spatial extent relative to a landscape of given size may maintain a quasi-equilibrium during the intervals between larger disturbances with a longer rotation period that punctuate the more stable periods. Such a system may be hierarchically nested as described earlier, with a complete set of equilibria punctuated at progressively longer intervals at larger scales (Figure 8.2). This is possible in forest landscapes in which high-severity disturbances occur rarely, so that several tree generations pass between disturbances and small gaps or other small-scale disturbances have a major impact on forest dynamics. Large landscapes may have a high probability of attaining quasi-equilibrium status in the northern hemlock–hardwood region, and the duration of such status is probably long. In contrast, on forest landscapes where large, high-severity disturbances occur at intervals that are about equal to tree longevity, disturbances continually disrupt the processes that would otherwise occur at all smaller spatial scales. Small areas, such as a stand, have a low probability of attaining quasi-equilibrium status, and the duration of such status is short when it does occur.

Stability of species composition

A hierarchy of forest change

Species composition change occurs at many different levels. Most ecologists agree that, to fully understand a process, one must bracket that process by looking at least one level higher and lower than the scale or magnitude of the process of interest. Throughout the book I have done this by putting stand dynamics in the middle of a three-part hierarchy of spatial scales: neighborhood, stand and landscape. Now, to understand magnitudes of compositional change, we need analogous brackets. The following four-level hierarchy starts with large magnitudes of compositional change and progresses to those of lower magnitude (Figure 8.3). It

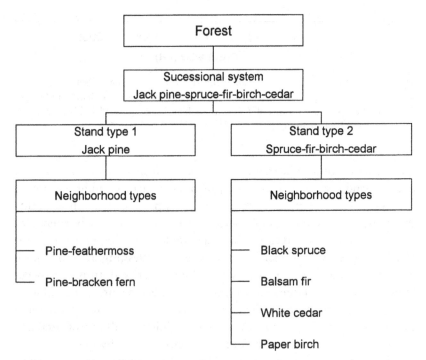

Figure 8.3. An example of the hierarchical levels of forest vegetation. Large differences in composition exist between forest and non-forest (not shown) and among successional systems, and relatively small differences exist among stand types and neighborhood types within a stand.

is best to view this whole hierarchy as occurring on a constant physiographic setting (with the same soil parent materials) and ask as you read: What changes would be necessary to convert among types or systems at each level? It is easy to predict differences in disturbance frequency and vegetation dynamics on different parent materials, or under different climates. It is not so easy to predict how changes will occur on uniform parent materials.

Forest versus non-forest
Changes of this very large magnitude often correspond to major climate changes. A climate that becomes too dry, too cold, or a fire frequency that is too high can convert·forest to non-forest. A variety of biologically caused switches may operate that enhance separation of forest and non-forest communities (Wilson and Agnew 1992, Zackrisson et al. 1997, Ponge et al. 1998).

Different forest successional systems (i.e. ecosystems) within a forested landscape
Moderately large climate changes that stay within the general envelope for existence of forest can cause switches in successional systems. For example, it is easy to view the near-boreal aspen–fir successional system as a colder version of the birch, white pine, hemlock–hardwood successional system. Such shifts among successional systems have been observed in the paleoecological record of Sylvania Wilderness Area, where an oak and pine system gave way to the modern birch, white pine, hemlock–hardwood system (Davis et al. 1998). Adjacent bodies of sandy and loamy parent material under the same climate could also very well result in an oak–pine successional system versus a hemlock–hardwood system, respectively.

Different stand types within a successional system
Here we clearly get into the realm of disturbances, since different stand types such as birch stands within a hemlock-forested landscape are created by disturbance. As we will see below, stand composition responds to disturbance severity, which is often correlated with disturbance type. The processes of succession and stand development also contribute to differentiation of stand types after major disturbance. A third major process here is neighborhood effects, which, as demonstrated in Chapter 6, can cause formation of patches that go beyond the neighborhood scale. The proportion of stands in different states within a successional system is driven by changes in disturbance frequency, which in the case of

fire and wind, are probably driven by changes in climate that are too small to cause an entirely new successional system or by human changes in the disturbance regime (e.g. Clark 1988, Clark *et al.* 1989, Clark and Royall 1995, Clark *et al.* 1996, Frelich 1995, Bergeron *et al.* 1998).

Note the difference between the sense of different stand types as used here (stands of varying composition due to differences in disturbance history) and the other commonly used sense, of adjacent stands on bodies of soil with different characteristics (i.e. pine on sandy soil next to maple on loamy soil), which, at least for the purposes of this book, would result in different successional systems or ecosystems.

Different neighborhood types within a stand

Neighborhood effects that operate at very small scales and gap dynamics (treefall and spot fire) are responsible for changes in neighborhood composition from one part of a stand to another. Examples include gaps in the hemlock–hardwood forest filled by yellow birch, clumps of basswood that originate from sprouts, and small groves of paper birch or red oak after spot fires. Gap size can be an important feature that determines which species enters a given gap (Runkle 1981, Peterson and Pickett 1995). Neighborhood differences may also result from different species composition in the shrub and herbaceous layers, in cases where the tree canopy composition remains the same.

The hierarchy and stand and landscape dynamics

The upper two levels of the hierarchy are mainly controlled by climate change of different magnitudes, or by different parent materials. Although these sorts of differences are worthy of the brief discussion above to help put things in context, they are generally beyond the scope of the book. If we assume a constant climate for a few thousand years, as was explained earlier in the chapter, then the successional system for a given physiographic setting will stay constant. That leaves compositional change in the bottom two levels of the hierarchy as the topic of interest. Composition at these two levels is influenced by disturbances and neighborhood effects, with the cause of neighborhood differences being the same as the cause of stand differences, but merely operating at a smaller scale. Now let's put these observations together in a logical fashion:

1. Stands in different successional stages make up a successional system.
2. Collections of adjacent stands make up a landscape.
3. Differences among stands and neighborhoods are caused by the type and severity of disturbances, and type of neighborhood effects.

Therefore, there must be some sort of explanatory relationship between neighborhood effects, disturbance severity, and stand and landscape dynamics. The remainder of this chapter examines the cross-scale linkages among neighborhood effects, stand-level change and landscape dynamics. In so doing, it draws from everything previously described in the book, and ties it together. This sequence consists of chapter sections that show: (1) how stands respond to disturbances of different severity; (2) how neighborhood effects fit into the picture; and (3) that (1) and (2) result in four logical categories of landscape dynamics.

Stand response to disturbance severity

Disturbance severity revisited: cumulative disturbance severity
Earlier, disturbance severity was described as the degree of mortality, of overstory and advanced regeneration, caused by a disturbance. To develop a conceptual model of stand response to disturbance over time, however, we need to make disturbance severity a dynamic concept, to account for the cumulative effects of repeated disturbance over time.

Cumulative disturbance severity is closely related to disturbance frequency: if two disturbances come at time intervals so close that the system does not recover between the disturbances, then the severities of the two disturbances may be totally or partially additive. The case study of hemlock–hardwood forest whereby stand-leveling winds are followed by fires within a few years, converting the forest to birch, is an example. Two disturbances of moderate severity, or several of low severity, occurring within a few years may sometimes have the same effect as one moderate or high-severity disturbance. For example, case studies of windstorm and cutting effects in the northern hardwoods show that several low-severity disturbances within 1–2 decades can create stands with similar size structure and composition as stands that had complete canopy removal at one time (Eyre and Zillgitt 1953, Frelich and Lorimer 1991a), although two moderate-severity disturbances, such as total canopy blowdown, obviously cannot occur twice within a short time.

The important factor that determines the degree to which successive episodes of disturbance are additive is whether the forest has sufficient time for recovery between disturbances. If the forest recovers to the pre-disturbance state after one disturbance, then a second disturbance of similar severity will not have any more impact than the first. Thus, any disturbance that results in a reduction in the number of mature trees and

advanced reproduction to a level lower than that after the previous disturbance, will have an additive effect to the previous disturbance. We can
define the term *cumulative disturbance severity* as:

> The running total of mortality among mature trees and advanced regen-
> eration over time, such that any positive difference between mortality
> caused by a new disturbance and recovery since the last disturbance, is
> added to the running total.

This cumulative form of the disturbance severity retains the same
problem with the earlier discrete form of the definition: that it is difficult
to reconcile mature trees and advanced regeneration into one scale. This
makes it tough to devise an accurate quantitative number for disturbance
severity, and therefore we will adhere to the three categories of disturbance used throughout the book: light cumulative disturbance means
that the cumulative effects of recent disturbances has only removed small
parts of the understory or overstory; moderate cumulative disturbance
severity means that one or more recent disturbances have removed most
of the understory or overstory; and high-severity cumulative disturbance
means that one or more recent disturbances have removed most the
understory and overstory.

Conceptual models of stand response

If a stand experiences a series of disturbances that gradually change in
severity, stand composition must necessarily change at some point. It is
impossible for any one species to be able to replace itself abundantly after
disturbances spanning the range of severities discussed in this book
(intense crown fires to single treefall gaps). Successional systems exist
wherein a shade-intolerant, early-successional species group reproduces
after severe disturbances, and a more shade-tolerant late-successional
species group reproduces after low-severity disturbance (Heinselman
1973, Grigal and Ohmann 1975, Frelich and Lorimer 1991a, Frelich and
Reich 1995, Frelich and Reich 1998). What, then, is the nature of the
relationship between stand composition (dominance by early- versus
late-successional species) and disturbance severity? Three possible models
have been proposed (Frelich and Reich 1998, Figure 8.4):

Model 1 – Continuous response. Change in disturbance severity leads to
a proportional change in species composition. Increasing or decreasing
the severity of successive disturbances slowly over time will lead to opposite pathways up and down the same slope.

Model 2 – Discontinuous response. Changing severity of disturbances
will lead to very little change until a threshold is reached, causing a

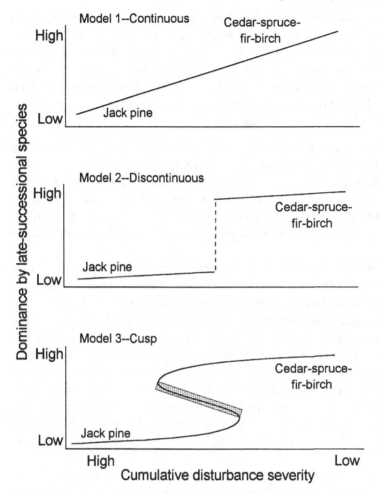

Figure 8.4. Three possible models for response of the forest – in this case composition of stands within the near-boreal successional system – to changing disturbance severity. After Frelich and Reich (1998).

sudden jump in composition if severity is increased or decreased a tiny amount at the threshold, which is the same going both directions.

Model 3 – Cusp response. Co-existence of two forest types with different composition is possible with moderate-severity cumulative disturbance, depending on the history of each stand. A large jump in composition can occur at the edge of the cusp going either direction, although the cusp is at a different point along the severity axis depending on direction. This model buffers the species composition against a considerable range of disturbance severities.

Keep in mind that the models show equilibrial attractors, analogous to most regression-defined relationships in community or ecosystem ecology, and do not specify the route(s) or length of time by which a given stand will arrive at a given point. There may be lags after disturbance, or a change in disturbance regime, before a given stand approaches the line/equilibrial attractor. It is possible to construct some general rules for these time-lag effects:

- When a stand has been experiencing a low cumulative disturbance severity and this suddenly switches to high severity, little time lag is likely, since severe disturbances remove disturbance-sensitive species and set the stage for immediate invasion of species adapted to high-severity disturbance. For model 1, the stand slides down the surface, and for the others, the stand goes over the threshold or cusp (Figure 8.5).
- When a stand has been experiencing a high cumulative disturbance severity and the regime suddenly changes to low-severity disturbances, the stand will be in a position beneath the line because disturbances less severe than those that allow establishment of a stand will not kill the trees. Successional processes will direct such a stand upward to the line (Figure 8.5). For example, a birch forest in which fires no longer occur would still be at the lowest level of the y-axis initially, but it could be replaced by a shade-tolerant species, either very slowly as the birch gradually die, or more rapidly, if a windstorm removes the birch and releases the understory, a process known as disturbance-mediated accelerated succession (Abrams and Scott 1989, Abrams and Nowacki 1992).
- Given the first two points, the response over time when there are major changes back-and-forth in cumulative severity, will exhibit hysteresis for all three models (Figure 8.5).

There is an interesting relationship between cumulative disturbance severity and the models of stand response with thresholds. For the cusp model, the two disturbance-severity thresholds can be viewed as lines with different slopes on a plot of cumulative amount of disturbance (y-axis) versus time (x-axis). Disturbance may occur at a high enough rate to maintain early-successional species (i.e. disturbance rate is higher than the lower, less steeply sloped line on Figure 8.6), or it may occur at a low enough rate to maintain late-successional species (i.e. rate is lower than the upper, more steeply sloped line on Figure 8.6A). Any time the trajectory of a stand catches up with one of these two lines, the cumulative dis-

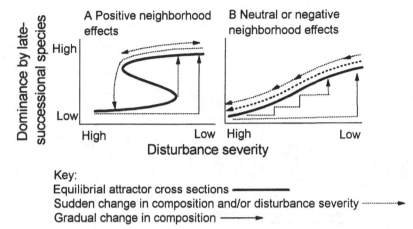

Key:
Equilibrial attractor cross sections ━━━━━━
Sudden change in composition and/or disturbance severity ┄┄┄┄►
Gradual change in composition ━━━►

Figure 8.5. Movements stands make in relation to the equilibrial attractor as disturbance regime severity changes over time. Dotted arrows represent sudden changes in composition in one jump, and solid arrows represent gradual changes in composition. A, model for stands with the cusp response. B, model for stands with the continuous response. Note hysteresis in response (i.e. increasing disturbance severity does not lead to the same route as decreasing disturbance severity in either case). After Frelich and Reich (1999).

turbance severity is beyond one of the cusp thresholds (see hypothetical examples of individual stand trajectories on Figure 8.6B). If the trajectory of a stand remains between the two lines, in the co-existence zone, then the stand composition will stay the same, whether currently early- or late-successional. For model 2, there is one line of intermediate slope on this same graph that would represent the threshold.

How to determine which model fits
With the basic conceptual models and their properties in place, and knowledge of the cumulative impacts of disturbance, one can proceed to collect evidence of which model directs stand movements among the states in a given successional system. A series of experiments where forests in various successional states are subjected to disturbances of different severities and then the response is observed would suffice. Immediately after the disturbances, the proportion of canopy trees, saplings, seedlings and seeds in the seedbank could be counted and compared with the pre-disturbance counts, thus quantifying the disturbance severity. After observing the recovery of the forest for each treatment, an assessment could be made as to the successional trajectory, and the

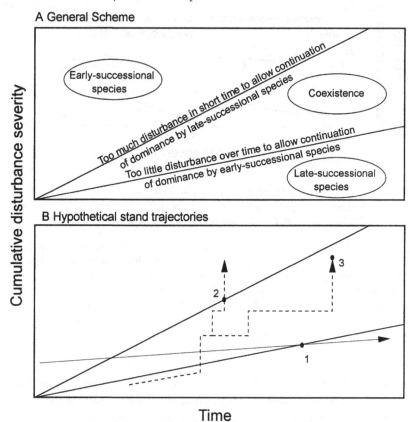

Figure 8.6. Cumulative disturbance over time and stand composition, or successional state, for the cusp model. A, general scheme with regions of high, moderate and low rates of disturbance. The different slopes of the two lines represent different rates of disturbance, which in turn represent threshold for minimum rate of disturbance necessary to maintain early successional species (lower line), or maximum rate of disturbance consistent with maintenance of late-successional species (upper line). B, trajectories of individual stands, illustrating the conditions under which stands will or will not convert to the alternate compositional state. The solid trajectory is a case where rate of disturbance is insufficient to maintain early successional species, leading to conversion to late-successional species at point 1. The dotted line trajectory illustrates the case where three disturbances in a short time are sufficient to cause conversion from late- to early-successional species at point 2. The same three disturbances distributed over a longer time, species conversion would not occur (point 3). After Frelich and Reich (1999).

mature post-disturbance stand composition could be plotted along an axis from low to high dominance by late-successional species versus observed disturbance severity. It would then be easy to see which of the three proposed models fits the data. Because trees are relatively long-lived, it would be necessary to follow post-disturbance composition for a few decades before we could be sure of the response. Unfortunately, studies with long-term observation that also carefully recorded the disturbance effects on the populations of trees and their propagules, and that also included a wide range of disturbance types and severities, are currently non-existent. Therefore, we will use retrospective studies of stand history, where the disturbances have been well documented, to examine stand response to low-, moderate-, and high-severity disturbance.

A synthesis of stand case studies from the Lake States documents that the cusp response (model 3) fits the relationship between hemlock–hardwood forests and aspen–paper birch forests (Frelich and Reich 1999). When the case studies are summarized in graphical form, the structure of the cusp is clearly visible (Figure 8.7). All of the 18 stands that experienced high-severity disturbances were heavily dominated by paper birch and aspen after disturbance, regardless of whether they were aspen–paper birch (7 cases) or hemlock–hardwood (11 cases) prior to disturbance. Case studies of response to moderate-severity disturbances showed that all stands remained very similar in composition after the disturbance as before, again regardless of the prior condition. Hemlock-hardwood stands and aspen–paper birch stands stayed similar in composition after two different types of disturbance: canopy clear cutting/heavy windthrow (hemlock–hardwood, 12 cases; aspen–paper birch, 7 cases), and surface fire (hemlock–hardwood, 10 cases; aspen–paper birch, 3 cases). Low-severity disturbances allowed hemlock–hardwood forest to stay in that condition (30 cases). Low-severity disturbance that continues over several decades also allows progressive conversion of aspen forest–paper birch to hemlock–hardwood (Heinselman 1954, not shown in Figure 8.7, but see Figure 8.5). The cusp relationship was also shown to occur between the jack pine stand type and the cedar–fir–spruce and birch near-boreal forest of northern Minnesota (Frelich and Reich 1998).

The two cases just cited contrast with white pine–birch forests in Minnesota, where there is a proportional relationship between disturbance severity and post-disturbance composition, with birch being progressively favored by disturbances of increasing severity (Heinselman 1973), so that model 1 is more appropriate.

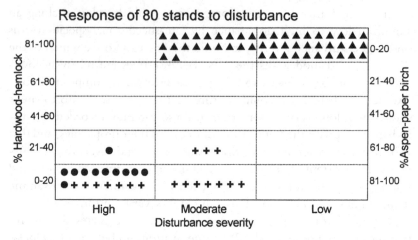

Figure 8.7. Case studies illustrating the cusp response in the field: stand response to disturbance in the aspen–birch–hemlock–hardwood successional system. The left and right *y*-axes represent canopy species composition (percent basal area or density) and are complementary to each other. Composition is shown in ordered categories after disturbance (3–30 years and 20–50 years post-disturbance for stands dominated by aspen–birch and hemlock–hardwood, respectively). Stands that were dominated by hemlock–hardwood before and after disturbance are indicated by triangles; those dominated by aspen–birch before and after disturbance are indicated by plus signs; and those dominated by hemlock–hardwood before disturbance and aspen–birch after disturbance are indicated by closed circles. After Frelich and Reich (1999); see that reference for sources of data.

The field of disturbance ecology is not advanced enough to allow a timely measurement of true severity other than the general categories of low, medium and high severity. Thus, relatively strong evidence of coexistence is necessary to show which model is appropriate, especially to prove the existence of the cusp in model 3 (Figure 8.4). For example, let us hypothesize that jack pine forests and cedar–spruce–fir–birch forests can both perpetuate themselves after moderately severe disturbance, with little or no intrusion of species from the opposite group. In that case, model 1 can be rejected. The co-existence of the two forest types after each receives a disturbance of precisely the same severity (within the moderate-severity range) would be evidence for the cusp response. If one were to observe a jack pine forest perpetuated after a surface fire and a cedar–spruce–fir–birch forest perpetuated after clear-cut logging (both disturbances defined as moderate above), it would be difficult to tell

whether there was a small difference in the severity of the two disturbances, such that surface fire was just to the left of the discontinuity in
model 2, and the logging just to the right of the discontinuity (Figure
8.4). Because we cannot insure that two different stands received disturbances of exactly the same severity, we need to show that substantial
overlap in disturbance severity within the middle portion of the range
perpetuates the alternate stand types. Therefore, perpetuation of two
alternate forest types by two different disturbance types, both within the
moderate-severity range, may be sufficiently strong evidence to distinguish the cusp response from the discontinuous type of response (Frelich
and Reich 1998).

Neighborhood effects: linking the three response models

The three models of response to changing disturbance severity can be
linked into a single three-dimensional surface that describes the response
of species composition (basically successional state) as a function of disturbance severity, and a second control variable that determines which of the
three basic shapes of response occurs. As it turns out, that second control
variable is the type of neighborhood effects exerted by the dominant
species in a forest stand. Stands with neutral–negative neighborhood
effects (such as white pine in the examples just discussed, which discou-
rage self-replacement or have a random chance of self-replacement) follow
the continuous response, while those with positive neighborhood effects
(likely to replace themselves, such as the hemlock–hardwood versus birch
example in Figure 8.7) follow the cusp response (Frelich *et al.* 1993,
Frelich and Reich 1995a,b, 1999). The resulting cusp-catastrophe model,
shown in general form in Figure 8.8, is one of a series of topological
models in mathematics known as catastrophe theory, which attempts to
explain the dynamics of systems in the physical and social sciences that
have both stable and unstable behaviors (Thom 1975, Zeeman 1976).
Jones (1977) and Holling (1981) were among the first to use the model to
describe long-term successional dynamics of forests, and applying it to
systems where insect infestations periodically kill the canopy in boreal
forests and where fire regulates the balance between forest and grassland.
The mathematical details of the model are given in a recent synthesis of the
subject by Ludwig *et al.* (1997). The response surface (Figure 8.8) shows
the four features that signify the cusp-catastrophe model (Poston and
Stewart 1978): (1) bimodality (above and below the cusp); (2) divergence

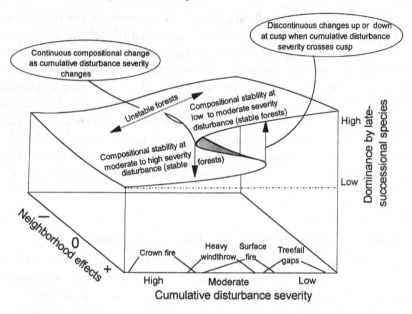

Figure 8.8. General form of the cusp-catastrophe model, showing how changing disturbance severity affects the composition of stands dominated by species with negative, neutral and positive neighborhood effects. After Frelich and Reich (1999).

in the neighborhood effect control variable; (3) hysteresis, or delayed response to changes in disturbance-regime severity in forests with positive neighborhood effects; and (4) sudden transitions caused by a small change in disturbance-regime severity at the edge of the cusp. These four properties manifest themselves topologically as the cusp.

Catastrophe theory is useful because it allows us to build a single descriptive, conceptual model for one or many forest ecosystem(s) that feature(s) continuous and discontinuous change. The theory is used to illustrate the neighborhood effect hypothesis of forest dynamics (Frelich and Reich 1995a, 1999):

> Forests with strong positive neighborhood effects should be stable with rare jumps in composition, while those with strong negative neighborhood effects should change continuously and unidirectionally, and those with neutral neighborhood effects appear unstable and 'wander' over time.

The major properties of this hypothesis of forest dynamics and the cusp-catastrophe model are consistent with those listed above under the three models of response (Frelich and Reich 1999):

1. Forest stands with positive neighborhood effects (along the front of the response surface; Figure 8.8) show considerable resistance to change in composition over a wide range of disturbance severities. When change does come, however, it is dramatic.

2. Forest stands dominated by species with neutral or negative neighborhood effects exhibit gradual, continuous change with changing disturbance severity. In a complementary fashion, all changes in cumulative disturbance severity cause a change in composition. The change may come all at once, if a forest that had been experiencing only low-severity disturbance suddenly is hit by a high-severity one, or in smaller steps.

3. The two control variables will sometimes change dramatically and/or abruptly, especially disturbance severity. Under these conditions, the response surface becomes an 'equilibrial attractor' (Poston and Stewart 1978). There may be a time lag in the response of the vegetation to a change in cumulative disturbance severity, the disturbance regime may not remain constant long enough for the vegetation to respond fully, and/or unique disturbance events may occur which are dramatically different in severity from those in the 'normal' regime. The position of a given stand may be off (above or below) the surface for variable periods of time, as described above for the three models of response.

4. Forest change has hysteresis for all forest types, again consistent with the discussion of response to the three models above.

5. It is impossible to predict the movements of any one stand across the response surface. We know that stands spend their time on or near the response surface, but not whether each one will stay in a certain place or where it will go next. If the disturbance regime remains stable, however, it may be possible to predict the proportion of stands on or near various parts of the surface.

Application of the cusp-catastrophe model

All that is necessary to apply the model in a semi-quantitative fashion for a given successional system is a knowledge of neighborhood effects and what types of disturbance would cause low-, moderate- and high-severity disturbance in the different stand types. This information has already been discussed in previous chapters for four case studies that I present here: (1) the paper birch, white pine, hemlock–hardwood successional system; (2) the near-boreal jack pine–aspen, spruce–fir–birch–cedar successional

Table 8.2. *Neighborhood effects and response to disturbance at the stand scale*

	Neighborhood effect type and direction[a]	
Stand type	Disturbance-activated	Overstory–understory
Near-boreal successional system		
Jack pine/black spruce	+	−
Aspen	+	−
Spruce–fir–birch	0	0
White cedar	−	+
Birch–pine–hemlock–hardwood successional system		
Paper birch	+	−
White pine	0	0
Hemlock–sugar maple	−	+
Spruce–fir–birch successional system		
Paper birch	+	−
Spruce–fir	−	+
Aspen–oak–sugar maple successional system		
Aspen	+	−
Red oak	0	0
Sugar maple	−	+/0

Note:
[a] Disturbance-activated neighborhood effects are a response to stand-killing fire, while overstory–understory effects are a response to canopy damage from windstorms. + indicates strong tendency for self-replacement, 0 indicates neutral chance of self-replacement, and − indicates a lack of self-replacement.

system; (3) the spruce–fir, birch successional system; and (4) the aspen–oak–sugar maple successional system of the 'Big Woods'. All of these successional systems have stand types with positive overstory–understory, positive disturbance-activated, neutral, and negative neighborhood effects (Table 8.2). Tying together information on previous response to disturbance from throughout the book, we also know the following general principles of forest response to cumulative disturbance effects:

I. Late-successional stands

- Disturbances that remove either the understory layer or the overstory layer will allow stand composition to remain the same.
- Disturbances that remove both the understory and overstory layers will allow replacement by early-successional species.

II. Early-successional stands

- Disturbances that remove the understory only will allow stand composition to remain the same.
- Disturbances that remove the overstory only may allow the stand to stay the same (for sprouters or serotinous seeders) or cause disturbance-mediated accelerated succession if an understory of shade-tolerant species is present.
- Disturbances that remove the understory and overstory will usually allow stand composition to remain the same.

III. Mid-successional stands

- Disturbances that remove the understory only will usually allow the formation of a mixed stand of the current species plus invaders after the disturbance.
- Disturbances that remove the overstory only will usually allow formation of a mixed stand of the current species plus late-successional invaders.
- Disturbances that remove the understory and overstory will allow replacement by early-successional species.

Disturbances that remove overstory, understory, or both are summarized in Table 8.3.

Near-boreal successional system
The dominant successional state under the historic natural disturbance regime is the jack pine and aspen forest type (Figure 8.9, state 1) that is perpetuated by moderate to high cumulative disturbance severity over time (Heinselman 1973, 1981a,b, Frelich and Reich 1995a,b). These forests have strong positive disturbance-activated neighborhood effects, in the form of serotinous seed rain and sprouting. In those stands missed by severe crown fire for more than a century, cumulative disturbance severity falls to a level that will no longer maintain early-successional species, and the stand becomes a jack pine and aspen stand with negative neighborhood effects, as the disturbance-activated effects fail to get switched on (Figure 8.9, state 2). Eventually, spruce, fir, cedar and paper birch attain places in the canopy under a low-severity regime of treefall gaps and neutral neighborhood effects (Figure 8.9, state 3). The stand may at first be beneath the response surface (an equilibrial attractor) and make its way up to state 3 gradually as succession proceeds. The stand will remain in state 3 as long as random replacement among the species

Table 8.3. *Disturbances and disturbance combinations that cause understory and/or overstory removal in Lake States forests*

Disturbances that take out both forest layers: the high-severity disturbance line up
  Crown fire
  Clear-cut and slash fire combination
  Clear-cut and heavy scarification combination
  Prolonged heavy deer browsing and stand-leveling wind combination
  Surface fire followed within a year by stand-leveling wind
  Stand-leveling wind followed by slash fire

Disturbances that take out the understory only (moderate severity)
  Herbivory by deer, moose, rabbits
  Exotic species invasion (European earthworms)
  Surface fire
  Scarification

Disturbances that take out the overstory only (moderate severity)
  Clear-cutting without ground disturbance (especially winter)
  Stand-leveling wind
  Insect infestation (e.g. budworm in mono-specific fir)
  Tree disease (in mono-specific stands)
  Independent crown fire (very rare)

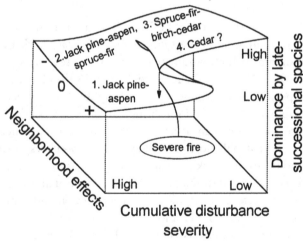

Figure 8.9. The cusp–catastrophe model applied to the near-boreal forest successional system. After Frelich and Reich (1999).

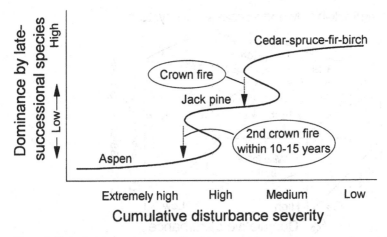

Figure 8.10. An alternative way of looking at stable states in the near-boreal forest: the three stable state model. The early-successional species are more finely divided into aspen (very early succession) and jack pine (early succession).

continues (Frelich and Reich 1995a). However, some have hypothesized that white cedar may eventually develop positive neighborhood effects and be the dominant of late-successional state 4 (Grigal and Ohmann 1975, Frelich and Reich 1995a,b, Figure 8.9). Crown fire, or any other disturbance combination listed in Table 8.3 as high severity, will send stands at states 2, 3, or 4 back to state 1, either down the slope or over the cusp. Disturbances listed as moderate severity (Table 8.3) should in theory maintain either state 1 or state 4.

The occurrence of two fires within a decade that eliminated jack pine from a stand and replaced it with aspen poses a logical challenge to this model. There are two ways to look at this. One is that jack pine and aspen are both 'early successional species', as the *y*-axis is labeled in Figure 8.9, and, therefore, the composition is 100% early successional whether either of these species dominates. The other way to look at the situation is to call aspen an earlier successional species than jack pine as it can withstand a higher cumulative disturbance severity than jack pine. If the stand burns twice in 10 years and the jack pine have no seed at the time of the second burn, then they have not recovered from the first fire. Thus, the two fires are additive in severity, so that the cumulative disturbance severity immediately after the second is 200%. To accommodate this, a model with three stable states would be necessary (Figure 8.10).

Hemlock-hardwood successional system

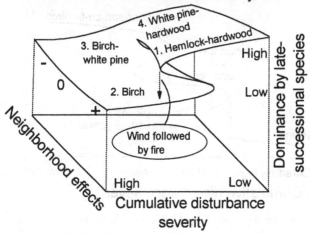

Figure 8.11. The cusp-catastrophe model applied to the hemlock–hardwood successional system. After Frelich and Reich (1999).

Birch, white pine, hemlock–hardwood successional system

The dominant state in this system under the historic natural disturbance regime is the low to moderate severity, positive neighborhood effects hemlock–sugar maple mixture (Figure 8.11, state 1). When a severe disturbance combination occurs (Table 8.3), the hemlock–hardwood is replaced by early-successional birch (Figure 8.11, state 2). If the disturbance regime does not change to one dominated by severe fires, then a stand will have two routes back to state 1. The first route is to pass through successional states 2, 3 and 4 – birch with negative neighborhood effects, birch with white pine, and white pine–hardwood, respectively – on its way back to state 1 (Figure 8.11). Surface fires or stand-leveling winds can push the white pine-dominated forests up and down the slope, so that they vary continuously from white pine mixed mostly with birch to white pine mixed mostly with hemlock and sugar maple. Once the fuel–fire feedback effect of hemlock and sugar maple gets established, however, the stands are strongly drawn back to state 1, since fire becomes very unlikely. The second route for a birch stand back to state 1 is directly up to it from beneath the response surface. Sometimes white pine seed sources are not available and that successional stage is skipped. If seed of hemlock and sugar maple is nearby, they can invade directly. If no fires occur for several decades and cumulative disturbance severity slips to a low level to the right of the cusp, birch may retain dominance with the potential to sprout if another fire occurs. Thus, the stand could retain positive disturbance-activated neighborhood effects even while it is being invaded by hemlock and sugar maple, so that it can stay on the positive neighborhood effect side through the whole successional sequence. Under such conditions, the stand would gradually approach state 1 from beneath, although it could also make a jump upwards if stand-leveling winds remove the birch canopy.

Spruce–fir–birch successional system

This system does not have any stand types with neutral or negative neighborhood effects, so that stands alternate between states 1 and 2 – above and below the cusp – and all succession takes place by replacement of birch and aspen by spruce and fir while the birch and aspen still retain the potential for positive disturbance-activated neighborhood effects (Figure 8.12). The contrast between this system, the near-boreal system, and the birch, white pine, hemlock–hardwood system is interesting. In this system the rotation period for severe fire is such that aspen and spruce–fir are always more or less equally represented on the landscape and there is no dominant state as there is in the other systems. The late-successional species, spruce and fir, do not change the fuel so that it is less flammable over time like late-succesional species in other systems (e.g. sugar maple and hemlock). Thus, there is no particular reason for stands to accumulate

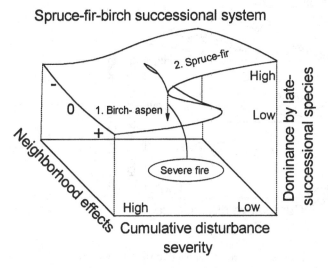

Figure 8.12. The cusp-catastrophe model applied to the near-boreal spruce-fir-birch successional system.

in the late-sucessional state. Instability in composition over time in this system is caused by lack of feedback between vegetation and disturbance severity, and by frequent renewal of the successional sequence by fire. Instability of white pine forests in the the birch, white pine, hemlock–hardwood successional system, on the other hand, is due to white pine's lack of positive neighborhood effects and sensitivity to changes in disturbance severity from one disturbance episode to the next.

Aspen–maple–oak (Big Woods) successional system
This system features alternate states of aspen forest stabilized by high cumulative disturbance severity and sugar maple–oak forest stabilized by low cumulative disturbance severity (Figure 8.13, states 1 and 4, respectively). The Big Woods are on the prairie–forest border, where surface fires are much more common than in hemlock–hardwood forests of Upper Michigan. The cumulative impact of these surface fires over a period of of 1–2 centuries regulates the movement of stands among the four states over time. If fire occurs only once or less per century then stands succeed to state 4 with sugar maple and some oak. More frequent fires – perhaps 2–3 per century – allow larger proportions of oak to mix with the maple (state 3). Very frequent fires remove sugar maple altogether and

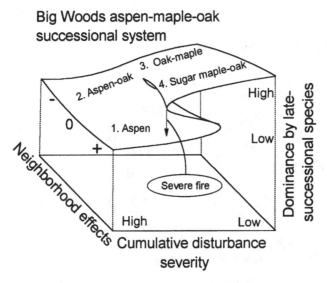

Figure 8.13. The cusp-catastrophe model applied to the 'Big Woods' aspen–maple–oak successional system.

create aspen or aspen–oak mixtures (states 1 and 2, respectively). The reason for including this system is to illustrate what happens when a species such as sugar maple loses its usual positive overstory–understory neighborhood effects. Surface fires are severe enough to kill seedlings and saplings of sugar maple, but not adults, leading to erratic self-replacement of sugar maple and co-existence with oaks that also have neutral neighborhood effects (Figure 8.13, state 3).

A classification of forest landscape dynamics

A cross-classification of disturbance and neighborhood effect interactions from the cusp-catastrophe models shows that there are four different types of forest landscape (Figure 8.14). Each of these has a characteristic set of dynamics. The landscape is still a collection of stands, and the category of landscape dynamics depends on which state within the successional system most stands reside at during a given period in time. For example, for the birch, white pine, hemlock–hardwood successional system under natural disturbance regime (Figure 8.11), all states in the successional system are represented at all times (assuming the landscape is large enough to support a quasi-equilibrium), but most stands reside in the hemlock and sugar maple dominated state 1.

Categories of landscape dynamics

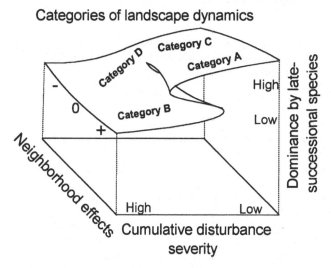

Figure 8.14.   Categories of landscape dynamics (where predominant stand type resides), as indicated by cross-classification of disturbance-regime severity and neighborhood effects. After Frelich and Reich (1999).

Table 8.4. *Characteristics of the four categories of dynamics*

Category A landscapes
(+ neighborhood effects and low disturbance severity)
- Adjacent stands of differing composition in areas with uniform soils are caused by species interactions
- Low–moderate-severity disturbance creates patches of different ages at neighborhood, stand and landscape scales
- Stability of age structure is low at the neighborhood scale, moderate at the stand scale and may be high at the landscape scale
- Compositional stability is high at neighborhood, stand and landscape spatial scales
- High-severity disturbance is rare and destabilizes composition at neighborhood, stand and landscape scales
- Successional episodes are initiated by rare high-severity disturbances
- The landscape consists of a matrix of late-successional species, all-aged neighborhoods and stands with a few stand-size inclusions of even-aged early-successional species

Category B landscapes
(+ neighborhood effects and high-severity disturbance)
- Patches of differing composition on uniform soils are caused by species interactions
- Low–moderate-severity disturbance plays a minor role in the overall dynamics
- High-severity disturbance is common and stabilizes composition at neighborhood, stand and landscape scales
- Stable age structure is rare at neighborhood and stand scales, but may occur on large landscapes
- Successional episodes are initiated by lack of crown fire
- Landscape is a matrix of large complex-shaped stands of even-aged early-successional species, with a few small stand-sized inclusions of uneven-aged, late-successional species

Category C landscapes
(0 or − neighborhood effects and low-severity disturbance regime)
- Patches of differing composition and differing age are caused by treefall gaps and other small disturbances
- Compositional stability is low at the neighborhood scale, but moderate-to-high at stand and landscape scales
- Stability of age structure is low at the neighborhood scale, but may be moderate-to-high at the stand and landscape scales
- Severe disturbance is rare and destabilizes composition at neighborhood, stand, and landscape scales
- Successional episodes at the stand scale are initiated by moderate to high-severity disturbance
- Landscape matrix is a fine-grained (neighborhood-scale) mosaic of mostly late-successional species, with stand-scale inclusions of even-aged early-successional species

Table 8.4 (*cont.*)

Category D landscapes
(0 or − neighborhood effects and moderate to high-severity disturbance)
- Patches of different composition caused by any disturbance
- Disturbances of any severity perpetuate instability of species composition
- Compositional stability is low at neighborhood and stand scales, but varies at the landscape scale
- Stability of age structure is low at neighborhood and stand scales, but may be high at landscape scales
- Successional episodes are initiated by disturbances of any severity
- Landscape matrix consists of large complex even-aged stands of mixed early and mid-successional species with stand-scale inclusions of old forest

---

Properties of the landscape dynamic categories

There are logical consequences of the different combinations of neighborhood effects and cumulative disturbance severities and some interesting contrasts among the four categories (Table 8.4). The major contrast is between those landscapes where most stands are dominated by species with strong positive neighborhood effects and those dominated by species with neutral to negative neighborhood effects. The most common natural disturbances in forests with positive neighborhood effects kill individual trees or the canopy of an entire stand without initiating succession. Composition remains stable after these disturbances, which create patches in different stages of development. Patches of differing species composition may be formed by processes other than disturbance, such as neighborhood effects and wave-form succession (Heinselman 1973, Frelich *et al.* 1993, Pacala *et al.* 1996). As long as the disturbance regime does not change, these landscapes covered by forests with positive neighborhood effects also tend to have stable species composition over time at neighborhood, stand and landscape scales.

Within the positive neighborhood-effect side of the response surface, the type of neighborhood effect provides the division between category A landscapes (dominated by stands with positive overstory–understory effects), and category B landscapes (dominated by stands with positive disturbance-activated effects). Severe disturbance causes a major change in composition ('compositional catastrophe'of Frelich and Reich 1999) and initiates an episode of succession in category A forests, while merely perpetuating the current species composition in category B forests. Thus, forested landscapes in category A will have a few stands dominated by

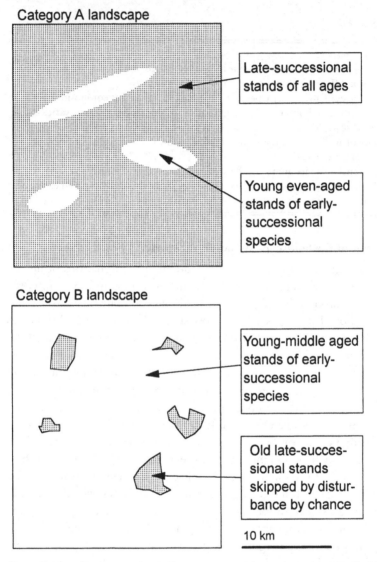

Figure 8.15. Contrasting patch characteristics of the category A and B landscapes.

early-successional species embedded in a stable matrix of shade-tolerant species (Bormann and Likens 1979), while category B forested land-scapes will have a matrix of young early-successional stands with a few embedded stands (or individual trees) of late-successional remnants that were by chance missed by disturbance (Heinselman 1973, Table 8.4, Figure 8.15).

Category C and D forest landscapes – in which most stands are dominated by trees with neutral to negative neighborhood effects – exhibit ongoing instability at some spatial scale, which depends on disturbance type. If disturbances are small and severity is low (e.g. treefall gaps), then individual neighborhoods will change composition frequently (every 1–2 tree lifetimes), while stands have a shifting-patch mosaic of different species that could be stable (Category C). Large moderately severe to severe disturbances on category D landscapes cause instability at neighborhood and stand scales, but there could be some stability at the landscape scale if the landscape meets criteria for a quasi-equilibrium (Table 8.4). On category C and D landscapes, patches of different ages and different species composition can be geographically coincidental rather than independent, as they were on forest landscapes with category A and B dynamics.

Remember that some landscapes, for example those forested by the spruce–fir–birch successional system, may not have a dominant successional state and therefore do not fit into one of the four categories. Instead, they have a hybrid system of dynamics or they may have alternating dynamics where the landscape shifts from one category to another over time.

Relationship of the landscape dynamic categories to previous forest concepts

Category A and C forest landscapes have been previously described as the 'shifting mosaic steady state' by Bormann and Likens (1979) and others. The original concept was developed to look at the theoretical steady state at the stand scale, in so-called all-aged stands of northern hardwoods that have a mosaic of gaps in different stages of recovery. These gaps could all be similar in composition (Bormann and Likens 1979), or also have some different species, depending on gap size (Fox 1977, Runkle 1981, Poulson and Platt 1996). Shifting mosaic steady state was also used as a multiple-scale concept for stands and landscapes (Shugart 1984). The refinement made with the new classification here is that those shifting mosaics with neutral neighborhood effects are separated into a shifting mosaic of ages and composition simultaneously, while those with positive neighborhood effects shift only in age structure.

Category B forest landscapes have been previously called non-climax systems (Loucks 1970). Several researchers discovered that canopy-killing disturbances can cause great changes in biomass and nutrient status of an ecosystem, but the species composition of the subsequent canopy remains similar to that before the disturbance (Dix and Swan 1971, Heinselman 1973, Johnson and Fryer 1989). In other words, there was a kind of stability in frequent disturbance. The existence of these forests was used by catastrophists as evidence for their side of the debate over climax versus non-climax forest that occurred during the 1970s and 1980s. Category A forests were the self-replacing climax systems dominated by shade-tolerant species of Clements (1936), Hough and Forbes (1943), and Lorimer (1977). The extension made here is that category A and B landscapes both have all successional states present, but that Category A landscapes are dominated by late-successional stands and category B landscapes are dominated by early-successional stands, and that feedbacks between the vegetation and disturbance could potentially maintain early and late-successional stands as alternate stable states.

Category D forests fit the classic 'individualistic succession' model or 'no stable endpoint' model, where continuous successional change due to frequent disturbance, climate change, or chance events such as local extinction and immigration prevents attainment of a stable equilibrium (Heinselman 1981b, Davis 1986, Foster and King 1986, Hubbell and Foster 1986). The extension made here is that these forests are intrinsically unstable at all spatial scales, and are yet another alternative within one successional system that includes stable forests, both climax and non-climax. Their existence does not rule out the possibility of stable forests.

Until now, there have been no simple ways of reconciling all the apparent empirical and conceptual discrepancies among forest researchers with regard to response to disturbance. When one realizes that response to disturbance may be linear or non-linear, and that this corresponds to the type of neighborhood effects, then it is logical that the cusp-catastrophe conceptual model can simultaneously encompass and incorporate these seemingly disparate patterns that previous research has put forth.

Occurrence of the landscape dynamic categories

A survey of the literature on forest dynamics reveals that many forest landscapes around the world fit (at least by my judgement) into the four categories of dynamics or alternating dynamics (Table 8.5). Some forest landscapes have also undergone one-time shifts from one dynamic category to another. If changes in disturbance frequency cause the predominant successional state in a successional system to switch, then the

Table 8.5. *Some temperate-zone forest types of the world and their dynamic categories*

Location	Forest type	Reference(s)
Category A forest landscapes		
Upper Michigan, USA	Hemlock–hardwood	Frelich and Lorimer 1991a, Davis *et al.* 1998
Eastern North America	Hemlock–hardwood	Runkle 1982, 1998
Massachusetts, USA	Hemlock–hardwood	Foster and Zebryk 1993
Pennsylvania, USA	Hemlock–hardwood	Peterson and Pickett 1995
Southeast Alaska, USA	Western hemlock/Sitka spruce	Deal *et al.* 1991
Central Japan	Japanese beech	Ohkubo *et al.* 1988
Japan	*Fagus crenata*	Yamamoto 1989
Monsoon Asia	Deciduous/beech	Nakashizuka and Iida 1995
Category B forest landscapes		
Saskatchewan, Canada	Boreal (jack pine)	Dix and Swan 1971
Minnesota, USA	Near-boreal (jack pine)	Heinselman 1973
Quebec, Canada	Near-boreal (red pine)	Bergeron and Dubuc 1989
Yellowstone National Park, USA	Lodgepole pine	Romme 1982, Turner *et al.* 1997
Southern Alberta, Canada	Lodgepole pine	Johnson and Fryer 1989
Category C forest landscapes		
Minnesota, USA	Spruce–fir–birch–cedar	Frelich and Reich 1995b
Northeastern USA	Red and white spruce/ paper birch	Pastor *et al.* 1987
New Zealand	*Nothofagus/Weinmannia*	Lusk and Smith 1998
Northeastern USA	Red and white spruce/ paper birch	Pastor *et al.* 1987
Category D forest landscapes		
Pennsylvania, USA	White pine	Hough and Forbes 1943
New England, USA	White pine	Foster 1988a,b
Minnesota, USA	White and red pine	Frelich and Reich 1995a
Northeastern USA	Red maple	Lorimer 1984
Eastern North America	Red oak	Lorimer 1983a, Abrams 1992
Northern Japan	Spruce–fir	Ishizuka *et al.* 1998
New Zealand	*Libocedrus bidwillii*	Veblen and Stewart 1982
Alternating landscapes		
Finland	Norway spruce–alder– birch aspen	Kuusela 1992
Minnesota, USA	Spruce–fir–birch ($A \rightleftharpoons B$)	Heinselman 1981b

category of landscape dynamics will also change. As previously mentioned, these changes could be driven by small-magnitude climatic change or by human interventions that change the natural disturbance regime. For example, the transition in the disturbance regime in most of the Lake States' hemlock–hardwood forest from wind-dominated to a regime dominated by logging followed by slash fire, accompanied a transition from a category A landscape where most stands were dominated by hemlock and sugar maple to a category B landscape where most stands were early-successional birch and aspen. This transition occurred between 1880 and 1920 in most of the northern Lake States region and adjacent Canada (Brisson *et al.* 1994) and is now partially reversing itself as the low- and moderate-severity disturbances of modern forestry – group selection cutting and clear cutting, respectively – have replaced the logging–slash fire combination.

Other transitions in landscape dynamics have been documented in the literature. A transition from a jack pine dominated landscape with category B dynamics to a spruce–fir–birch–cedar mixture with category C dynamics has followed fire exclusion in the near-boreal successional system (Frelich and Reich 1995a). The same transition (and vice versa) – caused by natural climate changes over the last few thousand years – has occurred in northern Minnesota and southern Quebec (Swain 1978, Bergeron *et al.*1998). The transitional sequence D–B–A, or D–A, has occurred when white-pine and birch dominated landscapes were logged, burned, and converted to birch–aspen forests, followed by fire suppression and succession to maple, or by direct fire suppression and succession to maple (Hough and Forbes 1943, Whitney 1987, Tester *et al.* 1997).

Categories of landscape dynamics and ecosystem type

In general, the more nutrient poor and coarser parent materials there are in an ecosystem, the easier it is to maintain early-successional stands and the harder it is to maintain late-successional stands. Thus, the cusp portion of the cusp-catastrophe conceptual model will fall farther towards the low end of the disturbance-severity gradient. For example, let us compare sugar maple versus aspen on two ecosystems: one with rich, loamy soil and one with poor, sandy soil. It has often been observed that sometimes aspen can be maintained by clear cutting, whereas on other sites clear cutting causes disturbance-mediated accelerated succession to sugar maple stands (Figure 8.16). The explanation for this lies in the fact that the drier the soil, the narrower the range of disturbance

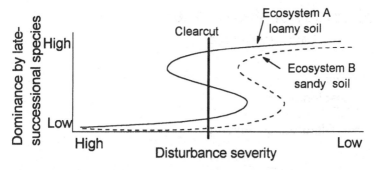

Figure 8.16. Changing ecosystem type may alter severity of disturbance for a given stand type, and the cusp may occur in a different location along the *x*-axis. In this example, a clear-cut in a stand on loamy soil may not be severe enough to convert the stand from late- to early-successional species, whereas the opposite may be true on sandy soil.

severities the maple can withstand without being converted to aspen. Conversely, the drier the soil, the wider the range of disturbance severities – especially toward the low end of the gradient – that can sustain aspen.

The general effect of moving from one ecosystem type to another for the conceptual model is that the cusp may appear at a different location along the disturbance-severity gradient. If one moves from a mesic to a dry-mesic soil type, not only will the level of aggressiveness of the late-successional species go down, but severe disturbance will become more likely. The forest will dry out enough to burn more often, and windfall slash will also be more likely to burn. These vegetation–soil–disturbance feedbacks are likely to cause a nutrient poor, dry ecosystem type to support more early-successional stands than late-successional ones, and the landscape dynamic category is more likely to be B than A.

Disturbance, neighborhood effects, herbivory and alternate stable states

It is common for ecologists to examine forests on adjacent bodies of soils that are different in texture, nutrient and water supply, and to project that disturbance and competition dynamics among the tree species there will lead to different successional trajectories. However, equally interesting is how and why alternate stable states develop on areas with relatively uniform soils and climate. Several concepts from the book can be united here to link alternate states caused by neighborhood effects and herbivory

The hierarchy of stable states

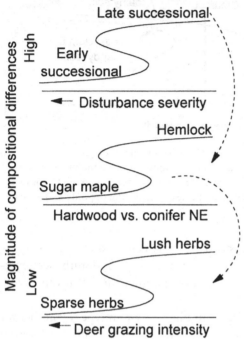

Figure 8.17. A hierarchy of alternate stable states with large, medium, and small compositional differences. Upper, early- versus late-successional species mediated by the interactions of disturbance severity and positive neighborhood effects. Middle, hemlock versus sugar maple, within the late-successional species group, mediated by hardwood versus conifer-type neighborhood effects. Lower, sugar maple with lush and sparse understories mediated by deer grazing.

by deer (from Chapters 6 and 7, respectively), with the alternate states caused by disturbance from this chapter. There is enough information for the birch–white pine, hemlock–hardwood successional system to examine the relationship among all three types of alternate states. Differences in disturbance regime severity create two alternate forest types that are very different from each other: an early-successional species group dominated by aspen and birch versus a late-successional species group dominated by sugar maple and hemlock. These two species groups are caused by a high- to moderate-severity disturbance regime versus a low- to moderate-severity disturbance regime, respectively (Figures 8.11, 8.17, 8.18). If we then take just the late-successional species group, separation into hemlock and sugar maple dominated stands is caused by conifer versus hardwood neighborhood effects (Table 6.5, Figures 6.18, 8.17, 8.19). These two alternate states are moderately different in species composition. Finally, if we examine just the sugar maple stand type, there are two alternative sub-stand types with lush and sparse understories, mediated in this case by deer grazing (Figures 7.4, 8.17, 8.20). The magnitude of species composition differences among these sugar maple types is relatively small.

Thus, the three types of alternate states presented in this book form a hierarchy of alternate states. States in the higher levels of the hierarchy can be split into alternate states with smaller magnitudes of compositional difference between them. For the birch–white pine, hemlock–hardwood successional system there are three such levels, each with a different factor that causes the split between alternate states: disturbance severity, neighborhood interactions between tree species, and interactions with mammalian herbivores. Another way to put it is that the biotic interactions with several different disturbance types creates multiple niches that are spatially separated. These niches can support different assemblages of species. Wilson and Agnew (1992) also show that these alternate states, caused by what they call 'switches', which seem to be pretty much synonymous with my neighborhood effect concept, also mediate many large-magnitude differences among adjacent vegetation types (forest versus non-forest) around the world. Also, these switches sharpen the transition between adjacent vegetation types that occur along gradual environmental gradients. Hence, alternate stable states can be created on a uniform environment or on an environmental gradient.

The self-organizing development of hierarchical structures that cause and maintain many alternate states should be an important research topic for two reasons. First, it represents a type of natural community key, as opposed to the artificial keys derived by those involved with classification of plant communities. One seldom encounters a community classification key that separates communities based on processes that formed them, such as grazed and ungrazed forests, or hemlock and sugar maple forests caused by neighborhood interactions as opposed to those caused by differences in disturbance history or underlying soil type. The hierarchy linked with community formation processes gives more powerful insight into structure than the currently used artificial community keys. Second, this type of hierarchy is an emergent property of our forest with disturbances, herbivores, and neighborhood interactions. When one also considers that all sorts of other as yet unknown alternate states probably also exist in our study area, one can see that these emergent structures

Figure 8.18. Alternate states with a large compositional difference: hemlock–hardwood forest (upper) versus paper birch (lower). Photos: Lee Frelich (upper); University of Minnesota Agricultural Experiment Station, Don Breneman (lower).

Figure 8.19. Alternate states with a medium compositional difference: hemlock forest (upper) versus sugar maple forest (lower). Photos: Lee Frelich.

Figure 8.20. Alternate states with a small compositional difference: sugar maple with a lush understory of wood nettle (Taylor's Woods, Hennepin County parks, Minnesota, upper) versus sugar maple forest with a sparse understory (Wood-Rill Scientific and Natural Area, Minnesota, lower). Photos: University of Minnesota Agricultural Experiment Station, Dave Hansen.

could significantly influence the formation and maintenance of diversity of life above that caused by variation in macroclimate and physiography. Future research should address the extent and influence of hierarchical systems of alternate states in community structure.

Conclusions

The cusp-catastrophe conceptual models appear to do a good job of qualitatively characterizing general patterns of forest development in response to disturbance for many temperate forests around the world. Assuming one is willing tentatively to accept the validity of the models, several lessons or implications follow:

1. The linear response to disturbance assumption. Many forest managers, and the models they use to project the future forest, assume a linear and stationary response of forest stands to disturbance. Some examples of these assumptions are that clear cutting will always lead to dominance by pioneer species; that future stands will regenerate to the same species as young stands on today's landscape; that pre-disturbance stand age dictates the post-disturbance response; and that post-harvest stand regeneration will always be a function of the pre-harvest species composition, unless planting is undertaken (e.g. Vanclay 1992, Jaakko Pöyry Consulting 1992, Peterson and Carson 1996, Ek et al. 1997). If forest managers are not aware of the likely correct model for response to disturbances of differing severities and of how disturbances vary in severity, surprises in stand response to manipulation may result. Sometimes they will expect a change in composition and get little or none, and other times large, unexpected changes will occur.

2. Severity of harvesting versus natural disturbances. Forest managers often assume that fires can be mimicked by logging. However, harvesting may be lower in severity than crown fires. Changing the severity of the disturbance regime may change forest composition, even if rotation periods and sizes of disturbances are purposely made the same as the natural disturbance regime. Several recent studies have shown that harvesting does not always mimic natural disturbance, precisely because of differences in severity of the disturbance (Ahlgren 1970, Noble et al. 1977, Abrams and Dickman 1982, Carleton and MacLellan 1994).

3. Punctuated stability exists at the stand scale. This is likely in forest stands dominated by species with strong positive neighborhood effects. In the words of Frelich and Reich (1995b): 'Stability never

lasts forever. The more stable a system is, the more severe a disturbance must be to overcome the stability and the less frequent such a disturbance will be. However, such a disturbance will eventually occur, changing the species composition and initiating an episode of succession.' Forests dominated by species with neutral or negative neighborhood effects, on the other hand, lack stability at some spatial scale. They undergo continuous change in response to disturbance, rather than periods of stability punctuated by sudden change.

4. Two types of instability exist at the stand scale. Stands on forest landscapes with category D dynamics are constantly changing because there is no buffering by the forest of the response to disturbances of differing severity, and severity is likely to change from one disturbance episode to the next. Stands on alternating landscapes are unstable because there is no feedback between the forest composition and the chance of severe disturbance.

5. Neighborhood effects link forest succession and stability across spatial scales from the individual tree to the landscape. Positive feedback loops for a species can lead to patchiness in forests and other vegetation types (Wilson and Agnew 1992). Neighborhood effects also help to determine the type of response to disturbance and influence the degree of stability the forest can possess.

6. Alternate stable states can exist that affect species composition at many levels: early- versus late-successional stand types; separation of different monospecific stand types within the early- or late-successional species groups; and separation of stand sub-types based on grazing and probably many other factors that may operate in the forest. These alternate stable states are more likely to occur when the dominant tree species have positive neighborhood effects, which often seem to be responsible for mediating alternate states. Hierarchies of alternate states show that many different natural communities can exist on relatively homogeneous sites.

7. Finally, many false debates over which models are the dominant ones have occupied the study of forest dynamics and disturbance regimes for several decades. These include continuous versus abrupt change, climax versus non–climax, stability versus instability, and succession versus differential growth rates. I hope to have shown in this book that these debates are caused by the lack of a complete picture possessed by investigators at various points along the way. Limited breadth of study sites with respect to different patterns of succession, short time horizons for studies, small size of study areas, and lack of spatial context

have all contributed to the fuzzy picture we have had of forest dynamics. The experience of many researchers and forest managers may be limited to one part of the proposed cusp-catastrophe model. At this point it appears that characterizing when, why and how often forests are stable or unstable will make the best sense of the many disparate observations of forest response to disturbance that have emerged over the last century.

References

Abbey, R.F., Jr. & Fujita, T.T. (1983). Tornadoes: the tornado outbreak of 3–4 April 1974. In *The Thunderstorm in Human Affairs*, ed. E. Kessler, pp. 37–66. Norman, Oklahoma: University of Oklahoma Press.

Abrams, M.D. (1992). Fire and the development of oak forests. *BioScience*, **42**, 346–353.

Abrams, M.D. & Dickman, D.E. (1982). Early revegetation of clear cut and burned jack pine sites in northern Lower Michigan. *Canadian Journal of Botany*, **60**, 946–954.

Abrams, M.D. & Nowacki, G.J. (1992). Historical variation in fire, oak recruitment, and post-logging accelerated succession in central Pennsylvania. *Bulletin Torrey Botanical Club*, **119**, 19–28.

Abrams, M.D. & Scott, M.L. (1989). Disturbance-mediated accelerated succession in two Michigan forest types. *Forest Science*, **35**, 42–49.

Agee, J.K. (1993). *Fire Ecology in Pacific Northwest Forests*. Washington, D.C.: Island Press.

Ahlgren, C.E. (1970). *Some Effects of Prescribed Burning on Jack Pine Reproduction in Northeastern Minnesota*. University of Minnesota Agricultural Experiment Station, Miscellaneous Report 94, Forestry Series 5–1970.

Ahlgren, C.E. & Ahlgren, I.F. (1981). Some effects of different forest litters on seed germination and growth. *Canadian Journal of Forest Research*, **11**, 710–714.

Alvarez, I.F., Rowney, D.L. & Cobb Jr., F.W. (1979). Mycorrhizae and growth of white fir seedlings in a mineral soil with and without organic layers in a California forest. *Canadian Journal of Forest Research*, **9**, 311–315.

Anderson, R.C. & Loucks, O.L. (1979). White-tail deer (*Odocoileus virginanus*) influence on structure and composition of *Tsuga canadensis* forests. *Journal of Applied Ecology*, **16**, 855–861.

Arno, S.F. & Sneck, K.M. (1977). *A Method for Determining Fire History in Coniferous Forests of the Mountain West*. U.S. Department of Agriculture, Forest Service, General Technical Report INT-42.

Augustine, D. & Frelich, L.E. (1998). White-tailed deer impacts on populations of an understory forb in fragmented deciduous forests. *Conservation Biology*, **12**, 995–1004.

Augustine, D., Frelich, L.E. & Jordan, P.A. (1998). Evidence for two alternative stable states in an ungulate grazing system. *Ecological Applications*, **8**, 1260–1269.

Banks, C.C. (1973). The strength of trees. *Journal of the Institute for Wood Science*, **6**, 44–50.

Beals, E.W., Cottam, G. & Vogl, R.J. (1960). Influence of deer on vegetation of the Apostle Islands, Wisconsin. *Journal of Wildlife Management*, **24**, 68–80.

Beatty, S.W. (1984). Influence of microtopography and canopy species on spatial patterns of forest understory plants. *Ecology*, **65**, 1406–1419.

Beatty, S.W. & Stone, E.L. (1986). The variety of soil microsites created by tree falls. *Canadian Journal of Forest Research*, **16**, 539–548.

Bergeron, Y. & Brisson, J. (1990). Fire regime in red pine stands at the northern limit of the species range. *Ecology*, **71**, 1352–1364.

Bergeron, Y. & Dubuc, M. (1989). Succession in the southern part of the Canadian boreal forest. *Vegetatio*, **79**, 51–63.

Bergeron, Y., Leduc, A., Morin, H. & Joyal, C. (1995). Balsam fir mortality following the last spruce budworm outbreak in northwestern Quebec. *Canadian Journal of Forest Research*, **25**, 1375–1384.

Bergeron, Y., Richard, P.J.H., Carcaillet, C., Gauthier, S., Flannigan, M. & Prairie, Y.T. (1998). Variability in fire frequency and forest composition in Canada's southeastern boreal forest: a challenge for sustainable forest management. *Conservation Ecology*, **2** [online], URL: http://www.consecol.org/vol2/iss2/art6.

Bessie, W.C. & Johnson, E.A. (1995). The relative importance of fuels and weather on fire behavior in subalpine forests. *Ecology*, **76**, 747–762.

Bicknell, S.H. (1982). Development of canopy stratification during early succession in northern hardwoods. *Forest Ecology and Management*, **4**, 41–51.

Blackburn, P. & Petty, J.A. (1988). Theoretical calculations of the influence of spacing on stand stability. *Forestry*, **61**, 235–244.

Blackburn, P., Petty, J.A. & Miller, K.F. (1988). An assessment of the static and dynamic factors involved in windthrow. *Forestry*, **61**, 29–43.

Blais, J.R. (1983). Trends in the frequency, extent and severity of spruce budworm outbreaks in eastern Canada. *Canadian Journal of Forest Research*, **13**, 539–532.

Boettcher, S.E. & Kalisz, P.J. (1990). Single-tree influence on soil properties in the mountains of eastern Kentucky. *Ecology*, **71**, 1365–1372.

Boose, E.R., Foster, D.R. & Fluet, M. (1994). Hurricane impacts to tropical and temperate forest landscapes. *Ecological Monographs*, **64**, 369–400.

Bormann, F.H. & Likens, G.E. (1979). *Pattern and Process in a Forested Ecosystem*. New York: Springer-Verlag.

Bossema, I. (1979). Jays and oaks: an eco-ethological study of a symbiosis. *Behavior*, **70**, 1–117.

Bourdo, E.A. (1956). A review of the General Land Office Survey and of its use in quantitative studies of former forests. *Ecology*, **37**, 754–768.

Brandner, T.A., Peterson, R.O. & Risenhoover, K.L. (1990). Balsam fir on Isle Royale: effects of moose herbivory and population density. *Ecology*, **71**, 155–164.

Braun, E.L. (1950). *Deciduous Forests of Eastern North America*. New York: Hafner.

Brisson, J., Bergeron, Y. Bouchard, A. & Leduc, A. (1994). Beech–maple dynamics in an old-growth forest in southern Quebec. *Ecoscience*, **1**, 40–46.

Bryson, R.A. (1966). Air masses, streamlines, and the boreal forest. *Geographical Bulletin*, **8**, 228–269.

Buckley, J.D. & Willis, E.H. (1970). ISOTOPES' radiocarbon measurements. VIII. *Radiocarbon*, **12**, 87–129.

Buell, M.F. & Niering, W.A. (1957). Fir–spruce–birch forest in northern Minnesota. *Ecology*, **38**, 602–610.

Buell, M.F., Buell, H.F. & Small, J.A. (1954). Fire in the history of Mettler's Woods. *Bulletin of the Torrey Botanical Club*, **81**, 253–255.

Calcote, R.R. (1995). Pollen source area and pollen productivity: evidence from forest hollows. *Journal of Ecology*, **83**, 591–602.

Calcote, R.R. (1998). Identifying forest stand types using pollen from forest hollows. *Holocene,* **8**, 423–432.

Candau, J-N., Fleming, R.A. & Hopkin, A. (1998). Spatiotemporal patterns of large-scale defoliation caused by the spruce budworm in Ontario since 1941. *Canadian Journal of Forest Research*, **28**, 1733–1741.

Canham, C.D. (1985). Suppression and release during canopy recruitment in *Acer saccharum*. *Bulletin of the Torrey Botanical Club*, **112**, 134–145.

Canham, C.D. (1989). Different responses to gaps among shade-tolerant tree species. *Ecology*, **70**, 548–550.

Canham, C.D. & Loucks, O.L. (1984). Catastrophic windthrow in the presettlement forests of Wisconsin. *Ecology*, **65**, 803–809.

Canham, C.D., Finzi, A.C., Pacala, S.W. & Burbank, D.H. (1994). Causes and consequences of resources heterogeneity in forests: interspecific variation in light transmission by canopy trees. *Canadian Journal of Forest Research*, **24**, 337–349.

Carleton, T.J., & MacLellan, P. (1994). Woody responses to fire versus clear-cutting logging: a comparative survey in the central Canadian boreal forest. *Ecoscience*, **1**, 141–152.

Cattelino, P.J., Noble, I.R., Slatyer, R.O. & Kessel, S.R. (1979). Predicting the multiple pathways of plant succession. *Environmental Management*, **3**, 41–50.

Channing, W. (1939). *New England Hurricanes*. Boston, Massachusetts: Walter Press.

Clark, J.S. (1988). Effects of climate change on fire regime in northwestern Minnesota. *Nature*, **334**, 233–235.

Clark, J.S. & Royall, P.D. (1995). Transformation of a northern hardwood forest by aboriginal (Iroquois) fire: charcoal evidence from Crawford Lake, Ontario, Canada. *The Holocene*, **5**, 1–9.

Clark, J.S., Merkt, J. & Müller, H. (1989). Post-glacial fire, vegetation, and human history on the northern alpine forelands, south-western Germany. *Journal of Ecology*, **77**, 897–925.

Clark, J.S., Royall, P.D. & Chumbley, C. (1996). The role of fire during climate change in an eastern deciduous forest at Devil's Bathtub, New York. *Ecology*, **77**, 2148–2166.

Clements, F.E. (1936). Nature and structure of the climax. *Journal of Ecology*, **24**, 252–284.

Cogbill, C.V. (1985). Dynamics of the boreal forests of the Laurentian Highlands, Canada. *Canadian Journal of Forest Research*, **15**, 252–261.

Cook, E.R. & Jacoby, C.G. (1977). Tree-ring–drought relationships in the Hudson Valley, New York. *Science*, **198**, 399–401.

Cornett, M.W., Reich, P.B. & Puettmann, K.J. (1997). Canopy feedbacks and microtopography regulate conifer seedling distribution in two Minnesota conifer–deciduous forests. *Ecoscience*, **4**, 353–364.

Cornett, M.W., Reich, P.B., Puettmann, K.J. & Frelich, L.E. (2000). Seedbed and moisture availability determine safe sites for early *Thuja occidentalis* (Cupressaceae) regeneration. *American Journal of Botany*, **87**, in press.

Court, A. (1974). The climate of the conterminous United States. In *Climates of North America,* ed. R.A. Bryson & K.F. Hare, pp. 193–266. New York: Elsevier.

Curtis, J.T. (1956). The modification of mid-latitude grasslands and forests by man. In *Man's Role in Changing the Face of the Earth*, ed. W.L. Thomas, pp. 721–736. Chicago, Illinois: University of Chicago Press.

Curtis, J.T. (1959). *The Vegetation of Wisconsin*. Madison, Wisconsin: The University of Wisconsin Press.

Cwynar, L.C. (1977). The recent fire history of Barron Township, Algonquin Park. *Canadian Journal of Botany*, **55**, 1524–1538.

Cwynar, L.C. (1978). Recent history of fire and vegetation from laminated sediment of Greenleaf Lake, Algonquin Park, Ontario. *Canadian Journal of Botany*, **56**, 10–21.

Czaplewski, R.L. & Reich, R.M. (1993). *Expected Value and Variance of Moran's Bivariate Spatial Autocorrelation Statistic for a Permutation Test.* U.S. Department of Agriculture, Forest Service, Research Paper RM-309.

Dahir, S.E. & Lorimer, C.G. (1996). Variation in canopy gap formation among developmental stages of northern hardwood stands. *Canadian Journal of Forest Research*, **26**, 1875–1892.

Dansereau, P-R. & Bergeron, Y. (1993). Fire history in the southern boreal forest of northwestern Quebec. *Canadian Journal of Forest Research*, **23**, 25–32.

Darley-Hill, S. & Johnson, W.C. (1981). Acorn disposal by the blue jay (*Cynocitta cristata*). *Oecologia*, **50**, 231–232.

Davis, M.B. (1981). Quaternary history and the stability of forest communities. In *Forest Succession: Concepts and Application*, ed. D. C. West, H.H. Shugart, & D.B. Botkin, pp. 132–153. New York: Springer-Verlag.

Davis, M.B. (1986). Climatic instability, time lags, and community disequilibrium. In *Community Ecology*, ed. J. Diamond, & T.J. Case, pp. 269–284. New York: Harper and Row.

Davis, M.B., Sugita, S., Calcote, R.R. & Frelich, L.E. (1992). Effects of invasion by *Tsuga canadensis* on a North American forest ecosystem. In *Responses of Forest Ecosystems to Environmental Changes*, ed. A. Teller, P. Mathy, & J.N.R. Jeffers, pp. 34–44. New York: Elsevier Applied Science.

Davis, M.B., Sugita, S., Calcote, R.R., Ferrari, J.B. & Frelich, L.E. (1994). Historical development of alternate communities in a hemlock–hardwood forest in Michigan, USA. In *Large-scale Ecology and Conservation Biology*, ed. R. May, N. Webb, & P. Edwards, pp. 19–39. Oxford: Blackwell.

Davis, M.B., Calcote, R.R., Sugita, S. & Takahara, H. (1998). Patchy invasion and the origin of a hemlock–hardwoods forest mosaic. *Ecology*, **79**, 2641–2659.

Deal, R.L., Oliver, C.D. & Bormann, B.T. (1991). Reconstruction of mixed hemlock-spruce stands in coastal Southeast Alaska. *Canadian Journal of Forest Research*, **21**, 643–654.

De Liocourt, F. (1898). De l'amenagement des sapinieres. *Societie Forestiere Franche-Comte Belfort Bulletin No. 6*, 369–405.

Deeming, J.E. & Brown, J.K. (1975). Fuel models in the national fire-danger rating system. *Journal of Forestry*, **73**, 347–350.

Dix, R.L. & Swan, J.M.A. (1971). The role of disturbance and succession in upland forest at Candle Lake, Saskatchewan. *Canadian Journal of Botany*, **49**, 657–676.

Dorr, J.A. & Eschman, D.F. (1970). *Geology of Michigan.* Ann Arbor, Michigan: University of Michigan Press.

Doswell, C.A. (1980). Synoptic-scale environments associated with high plains severe thunderstorms. *Bulletin of the American Meteorological Society*, **61**, 1388–1400.

Downs, A.A. (1946). Response to release of sugar maple, white oak, and yellow poplar. *Journal of Forestry*, **44**, 22–27.

Drury, W.H. & Nisbet, I.C.T. (1973). Succession. *Arnold Arboretum Journal*, **54**, 331–368.

Dublin, H.S., Sinclair, A.R.E. & McGlade, J. (1990). Elephants and fire as causes of multiple stable states in the Serengeti-Mara woodlands. *Journal of Animal Ecology*, **59**, 1147–1164.

Dunn, C.P., Guntenspergen, G.R. & Dorney, J.R. (1983). Catastrophic wind disturbance in an old-growth hemlock–hardwood forest, Wisconsin. *Canadian Journal of Botany*, **61**, 211–217.

Dynesius, M. & Jonsson, B.G. (1991). Dating uprooted trees: comparison and application of eight methods in a boreal forest. *Canadian Journal of Forest Research*, **21**, 655–665.

Eagleman, J.R., Muirhead, V.U. & Willems, N. (1975). *Thunderstorms, Tornadoes, and Building Damage.* Lexington, Massachusetts: D.C. Heath and Company.

Eichenlaub, V.L. (1979). *Weather and Climate of the Great Lakes Region.* Notre Dame, Indiana: University of Notre Dame Press.

Ek, A.R., Robinson, A.P., Radtke, P.R. & Walters, D.K. (1997). Development and testing of regeneration imputation models for forests in Minnesota. *Forest Ecology and Management*, **94**, 129–140.

Eyre, F.H. & Zilligitt, W.M. (1953). *Partial Cuttings in the Northern Hardwoods of the Lake States.* Technical Bulletin 1076. Washington, D.C.: U.S. Department of Agriculture.

Facelli, J.M. & Pickett, S.T.A. (1990). Markovian chains and the role of history in succession. *Trends in Ecology and Evolution*, **5**, 27–30.

Ferarri, J.B. (1999). Fine-scale patterns of leaf litterfall and nitrogen cycling in an old-growth forest. *Canadian Journal of Forest Research*, **29**, 291–302.

Finzi, A.C., Canham, C.D. & Van Breemen, N. (1998). Canopy tree soil interactions within temperate forests: species effects on pH and cations. *Ecological Applications*, **8**, 447–454

Foster, D.R. (1988a). Disturbance history, community organization, and vegetation dynamics of the old-growth Pisgah Forest, southwestern New Hampshire, USA. *Journal of Ecology*, **76**, 105–134.

Foster, D.R. (1988b). Species and stand response to catastrophic wind in central New England, USA. *Journal of Ecology*, **76**, 135–151.

Foster, D.R. & King, G.A. (1986). Vegetation pattern and diversity in S.E. Labrador,

Canada: *Betula papyrifera* (birch) forest development in relation to fire history and physiography. *Journal of Ecology*, **74**, 465–483.

Foster, D.R. & Zebryk, T.M. (1993). Long-term vegetation dynamics and distur-bance history of a *Tsuga*-dominated forest in New England. *Ecology*, **74**, 982–998.

Fox, J.F. (1977). Alternation and coexistence of tree species. *American Naturalist*, **111**, 69–89.

Franklin, J.F. & Hemstrom, M.A. (1981). Aspects of succession in the coniferous forests of the Pacific Northwest. In *Forest Succession: Concepts and Application*, ed. D.C. West, H.H. Shugart, & D.B. Botkin, pp. 212–229. New York: Springer-Verlag.

Fraser, A.I. (1962). The soil and roots as factors in tree stability. *Forestry*, **35**, 117–127.

Frelich, L.E. (1986). *Natural disturbance frequencies in the hemlock–hardwood forests of the Upper Great Lakes Region*. Ph.D. Thesis, University of Wisconsin-Madison.

Frelich, L.E. (1992). The relationship of natural disturbances to white pine stand development. In *White Pine Symposium Proceedings: History, Ecology, Policy and Management*, ed. R.A. Stine, & M.J. Baughman, pp. 27–37. St Paul, Minnesota: Department of Forest Resources, College of Natural Resources and Minnesota Extension Service.

Frelich, L.E. (1995). Old forest in the Lake States today and before European settle-ment. *Natural Areas Journal*, **15**, 157–167.

Frelich, L.E. & Graumlich, L.J. (1994). Age class distribution and spatial patterns in an old-growth hemlock–hardwood forest. *Canadian Journal of Forest Research*, **24**, 1939–1947.

Frelich, L.E. & Lorimer, C.G. (1985). Current and predicted long-term effects of deer browsing in hemlock forests in Michigan, USA. *Biological Conservation*, **34**, 99–120.

Frelich, L.E. & Lorimer, C.G. (1991a). Natural disturbance regimes in hemlock–hardwood forests of the Upper Great Lakes Region. *Ecological Monographs*, **61**, 145–164.

Frelich, L.E. & Lorimer, C.G. (1991b). A simulation of landscape dynamics in old-growth northern hardwood forests. *Journal of Ecology*, **79**, 223–233.

Frelich, L.E. & Martin, G.L. (1988). Effects of crown expansion into gaps on evalua-tion of disturbance intensity in northern hardwood forests. *Forest Science*, **34**, 530–536.

Frelich, L.E. & Reich, P.B. (1995a). Spatial patterns and succession in a Minnesota southern-boreal forest. *Ecological Monographs*, **65**, 325–346.

Frelich, L.E. & Reich, P.B. (1995b). Neighborhood effects, disturbance, and succes-sion in forests of the Western Great Lakes Region. *Ecoscience*, **2**, 148–158.

Frelich, L.E. & Reich, P.B. (1998). Disturbance severity and threshold responses in the boreal forest. *Conservation Ecology*, 2[online], URL: http://www.conse-col.org/vol2/iss2/art7.

Frelich, L.E. & Reich, P.B. (1999). Neighborhood effects, disturbance severity, and community stability in forests. *Ecosystems*, **2**, 151–166.

Frelich, L.E., Calcote, R.R., Davis, M.B. & Pastor, J. (1993). Patch formation and maintenance in an old growth hemlock–hardwood forest. *Ecology*, **74**, 513–527.

Frelich, L.E., Reich, P.B. & Haight, R. (1998a). Wind disturbance and conservation of old-growth forest remnants in the Upper Midwest. *Ecological Society of America 1998 Annual Meeting Abstracts*, p. 167.

Frelich, L.E., Reich, P.B., Sugita, S., Davis, M.B. & Friedman, S.K. (1998b). Neighborhood effects in forests: implications for within stand patch structure and management. *Journal of Ecology*, **86**, 149–161.

Frissell, S.S. (1973). The importance of fire as a natural ecological factor in Itasca State Park, Minnesota. *Quaternary Research*, **3**, 397–407.

Fritts, H.C. (1976). *Tree Rings and Climate*. New York: Academic Press.

Fujita, T.T. (1978). *Manual of Downburst Identification for Project NIMROD*. Satellite and mesometeorology research project Research Paper No. 156. Chicago, Illinois: University of Chicago Department of Geophysical Sciences.

Gates, F.C. & Nichols, G.E. (1930). Relation between age and diameter in trees of the primeval northern hardwood forest. *Journal of Forestry*, **28**, 395–398.

Gleason, H.A. (1927). Further views on the succession concept. *Ecology*, **8**, 299–326.

Glitzenstein, J.S., Harcombe, P.A. & Streng, D.A. (1986). Disturbance, succession, and maintenance of species diversity in an East Texas forest. *Ecological Monographs*, **56**, 243–258.

Gloyne, R.W. (1968). The structure of wind and its relevance to forestry. *Forestry*, Supplement, 7–19.

Godman, R.M. (1968). Culture of young stands. In *Proceedings of the Sugar Maple Conference*, pp. 82–87. Houghton, Michigan: Michigan Technological University.

Godman, R.M. & Marquis, D.A. (1969). Thinning and pruning in young birch stands. In *Proceedings of the Birch Symposium*, pp. 119–127. U.S. Department of Agriculture, Forest Service, Northeast Forest Experiment Station.

Graham, S.A. (1941a). The question of hemlock establishment. *Journal of Forestry*, **39**, 567–569.

Graham, S.A. (1941b). Climax forests of the Upper Peninsula of Michigan. *Ecology*, **22**, 355–362.

Grigal, D.F. & Ohmann, L.F. (1975). Classification, description, and dynamics of upland plant communities within a Minnesota wilderness area. *Ecological Monographs*, **45**, 389–407.

Grimm, E.C. (1984). Fire and other factors controlling the Big Woods vegetation of Minnesota in the mid-nineteenth century. *Ecological Monographs*, **54**, 291–311.

Groot, A. & Horton, B.J. (1994). Age and size structure of natural and second-growth peatland *Picea mariana* stands. *Canadian Journal of Forest Research*, **24**, 225–233.

Haines, D.A. & Sando, R.W. (1969). *Climatic Conditions Preceding Historically Great Fires in the North Central Region*. Research Paper NC-34. Washington, D.C.: U.S. Department of Agriculture, Forest Service.

Hardy, Y., Mainville, M. & Schmitt, D.M. (1980). *An Atlas of Spruce Budworm Defoliation in Eastern North America, 1938–80*. U.S. Department of Agriculture, Forest Service, Miscellaneous Publication No. 1449.

He, H.S. & Mladenoff, D.J. (1999). Spatially explicit and stochastic simulation of forest-landscape fire disturbance and succession. *Ecology*, **80**, 81–99.

Heinselman, M.L. (1954). The extent of natural conversion to other species in the Lake States aspen–birch type. *Journal of Forestry*, **52**, 737–738.

Heinselman, M.L. (1963). Forest sites, bog processes, and peatland types in the glacial Lake Agassiz region, Minnesota. *Ecological Monographs*, **33**, 327–374.

Heinselman, M.L. (1970). Landscape evolution, peatland types, and the environment in the glacial Lake Agassiz Peatlands Natural Area, Minnesota. *Ecological Monographs*, **40**, 235–261.

Heinselman, M.L. (1973). Fire in the virgin forests of the Boundary Waters Canoe Area, Minnesota. *Quaternary Research*, **3**, 329–382.

Heinselman, M.L. (1981a). Fire intensity and frequency as factors in the distribution and structure of northern ecosystems. In *Fire Regimes and Ecosystem Properties*, U.S. Department of Agriculture, Forest Service, Technical Report WO-26, pp. 7–57.

Heinselman, M.L. (1981b). Fire and succession in the conifer forests of northern North America. In *Forest Succession: Concepts and Applications*, ed. D.C. West, H.H. Shugart, & D.B. Botkin, pp. 374–405. New York: Springer-Verlag.

Heinselman, M.L. (1996). *The Boundary Waters Wilderness Ecosystem*. Minneapolis, Minnesota: University of Minnesota Press.

Henry, J.D. & Swan, J.M.A. (1974). Reconstruction of forest history from live and dead plant material: an approach to the study of forest succession in southwest New Hampshire. *Ecology*, **55**, 772–783.

Hibbs, D.E. (1982). Gap dynamics in a hemlock–hardwood forest. *Canadian Journal of Forest Research*, **12**, 522–527.

Holling, C.S. (1981). Forest insects, forest fires, and resilience. In *Fire Regimes and Ecosystem Properties*, U.S. Department of Agriculture, Forest Service, Technical Report WO-26, pp. 445–464.

Hoppes, W.G. (1988). Seedfall patterns of several species of bird-dispersed plants in an Illinois woodland. *Ecology*, **69**, 320–329.

Horn, H. (1974). The ecology of secondary succession. *Annual Review of Ecology and Systematics*, **5**, 25–37.

Hough, A.F. (1932). Some diameter distributions in forest stands of northwestern Pennsylvania. *Journal of Forestry*, **30**, 933–943.

Hough, A.F. & Forbes, R.D. (1943). The ecology and silvics of forests in the high plateaus of Pennsylvania. *Ecological Monographs*, **13**, 299–320.

Hough, J.L. (1958). *Geology of the Great Lakes*. Urbana, Illinois: University of Illinois Press.

Hubble, S.P. & Foster, R.B. (1986). Biology, chance, and history, and the structure of tropical rain forest tree communities. In *Community Ecology*, ed. J. Diamond, & T.J. Case, pp. 314–329. New York: Harper and Row.

Irving, R.D. (1880). Geology of the eastern Lake Superior district. *Geology of Wisconsin*, **3**, 89–91.

Ishizuka, M., Toyooka, H., Osawa, A., Kushima, H., Kanazawa, Y. & Sato, A. (1998). Secondary succession following catastrophic windthrow in a boreal forest in Hokkaido, Japan: the timing of tree establishment. *Journal of Sustainable Forestry*, **6**, 367–388.

Jaakko Pöyry Consulting (1992). *Maintaining Forest Productivity and the Forest Resource base. A technical paper for the Generic Environmental Impact Statement on Timber Harvesting and Forest Management in Minnesota*. Tarrytown, New York: Jaakko Pöyry Consulting.

Jacobson, G.L. & Grimm, E.C. (1986). A numerical analysis of Holocene forest and prairie vegetation in central Minnesota. *Ecology*, **67**, 958–966.

Johnson, E.A. (1992). *Fire and Vegetation Dynamics: Studies from the North American Boreal Forest*. Cambridge: Cambridge University Press.

Johnson, E.A. & Fryer, G.I. (1989). Population dynamics in lodgepole pine–Engelmann spruce forests. *Ecology*, **70**, 1335–1345.

Johnson, E.A. & Gutsell, S.L. (1994). Fire frequency models, methods and interpretations. *Advances in Ecological Research*, **25**, 239–287.

Johnson, E.A. & Larsen, C.P.S. (1991). Climatically induced change in fire frequency in the southern Canadian Rockies. *Ecology*, **72**, 192–201.

Johnson, E.A. & Wowchuk, D.R. (1993). Wildfires in the southern Canadian Rocky Mountains and their relationship to mid-tropospheric anomalies. *Canadian Journal of Forest Research*, **23**, 1213–1222.

Jones, D.D. (1977). The application of catastrophe theory to ecological systems. In *New Directions in the Analysis of Ecological Systems, part 2*, ed. G.S. Innis, pp. 133–148. LaJolla, California: Simulation Council.

Jordan, J.K. (1973). *A Soil Resource Inventory of Sylvania Recreation Area*. Watersmeet, Michigan: U.S. Department of Agriculture, Forest Service, Ottawa National Forest.

Kilgore, B.M. (1981). Fire in ecosystem distribution and structure: western forests and scrublands. In *Fire Regimes and Ecosystem Properties*, U.S. Department of Agriculture, Forest Service, Technical Report WO-26, pp. 58–89.

King, D.A. (1986). Tree form, height growth, and susceptibility to wind damage in *Acer saccharum*. *Ecology*, **67**, 980–990.

Klein, W.H. (1957). *Principal Tracks and Mean Frequencies of Cyclones and Anticyclones in the Northern Hemisphere*. Research Paper No. 40, U.S. Weather Bureau, U.S. Government Printing Office.

Knowles, P. & Grant, M.C. (1983). Age and size structure analysis of Engelmann spruce, ponderosa pine, lodgepole pine, and limber pine in Colorado. *Ecology*, **64**, 1–9.

Kronzucker, H.J., Siddiqi, M.Y. & Glass, A.D.M. (1997). Conifer root discrimination against soil nitrate and the ecology of forest succession. *Nature*, **385**, 59–61.

Kuusela, K. (1992). Boreal forestry in Finland: a fire ecology without fire. *Unasylva*, **43** (170), 22–25.

Leak, W.B. (1975). Age distribution in virgin red spruce and northern hardwoods. *Ecology*, **56**, 1451–1454.

Legendre, P. & Fortin, J-M. (1989). Spatial pattern and ecological analysis. *Vegetatio*, **80**, 107–138.

Leitner, L.A., Dunn, C.P., Gunstenspergen, G.R., Stearns, F. & Sharpe, D.M. (1991). Effects of site, landscape features, and fire regime on vegetation patterns in presettlement southern Wisconsin. *Landscape Ecology*, **5**, 203–217.

Lertzman, K. & Fall, J. (1998). From forest stands to landscapes: spatial scales and the roles of disturbances. In *Ecological Scale: Theory and Applications*, ed. D.L. Peterson, & V.T. Parker, pp. 339–367. New York: Columbia University Press.

Leopold, A. (1943). Deer irruptions. *Wisconsin Conservation Bulletin*, **8**, 3–11.

Lippe, E., De Smidt, J.T. & Glenn-Lewin, D.C. (1985). Markov models and succession: a test from a heathland in the Netherlands. *Journal of Ecology*, **73**, 775–791.

Lorimer, C.G. (1977). The presettlement forest and natural disturbance cycle of northeastern Maine. *Ecology*, **58**, 139–148.

Lorimer, C.G. (1980). Age structure and disturbance history of a southern Appalachian virgin forest. *Ecology*, **61**, 1169–1184.

Lorimer, C.G. (1981). The use of land survey records in estimating presettlement fire frequency. In *Proceedings of the Fire History Workshop*, General Technical Report RM-81, pp. 57–62.

Lorimer, C.G. (1983a). Eighty-year development of northern red oak after partial cutting in a mixed-species Wisconsin forest. *Forest Science*, **29**, 371–383.

Lorimer, C.G. (1983b). Tests of age-independent competition indices for individual trees in natural hardwood stands. *Forest Ecology and Management*, **6**, 343–360.

Lorimer, C.G. (1984). Development of red maple understory in northeastern oak forests. *Forest Science*, **30**, 3–22.

Lorimer, C.G. (1985). Methodological considerations in the analysis of forest disturbance history. *Canadian Journal of Forest Research*, **15**, 200–213.

Lorimer, C.G. & Frelich, L.E. (1984). A simulation of equilibrium diameter distributions of sugar maple (*Acer saccharum*). *Bulletin of the Torrey Botanical Club*, **111**, 193–199.

Lorimer, C.G. & Frelich, L.E. (1989). A methodology for estimating canopy disturbance frequency and intensity in dense temperate forests. *Canadian Journal of Forest Research*, **19**, 651–663.

Lorimer, C.G. & Frelich, L.E. (1998). A structural alternative to chronosequence analysis for uneven-aged northern hardwood forests. *Journal of Sustainable Forestry*, **6**, 347–366.

Lorimer, C.G. & Gough, W.R. (1988). Frequency of drought and severe fire weather in north-eastern Wisconsin. *Journal of Environmental Management*, **26**, 203–219.

Lorimer, C.G. & Krug, A.G. (1983). Diameter distributions in even-aged stands of shade-tolerant and mid-tolerant trees species. *American Midland Naturalist*, **109**, 331–345.

Lorimer, C.G., Frelich, L.E. & Nordheim, E.V. (1988). Estimating gap origin probabilities for canopy trees. *Ecology*, **69**, 778–785.

Loucks, O.L. (1970). Evolution of diversity, efficiency, and community stability. *American Zoologist*, **10**, 17–25.

Ludwig, D., Walker, B. & Holling, C.S. (1997). Sustainability, stability, and resilience. *Conservation Ecology*, 1[online], URL: http://www.consecol.org/vol1/iss1/art7.

Lusk, C.H. & Smith, B. (1998). Life history differences and tree species coexistence in an old-growth New Zealand rainforest. *Ecology*, **79**, 795–806.

Lyon, C.J. (1935). Rainfall and hemlock growth in New Hampshire. *Journal of Forestry*, **33**, 162–168.

Lyon, C.J. (1936). Tree ring width as an index of physiological dryness in New England. *Ecology*, **17**, 457–478.

Madany, M.H., Swetnam, T.W. & West, N.E. (1982). Comparison of two approaches for determining fire dates from tree scars. *Forest Science*, **28**, 856–861.

Maissurow, D.K. (1935). Fire as a necessary factor in the perpetuation of white pine. *Journal of Forestry*, **33**, 373–378.

Maissurow, D.K. (1941). The role of fire in the perpetuation of virgin forests of northern Wisconsin. *Journal of Forestry*, **39**, 201–207.

Marschner, F.J. (1975). *The Original Vegetation of Minnesota* (map). Ed. M.L. Heinselman, St. Paul, Minnesota: U.S. Department of Agriculture, Forest Service, North Central Forest Experiment Station.

Marshall, R. (1927). *The Growth of Hemlock Before and After Release from Suppression.* Petersham, Mass.: Harvard Forest Bulletin No. 11.

May, R.M. (1977). Thresholds and breakpoints in ecosystems with a multiplicity of stable states. *Nature*, **269**, 471–477.

McCullough, D.G., Werner, R.A. & Neumann, D. (1998). Fire and insects in northern and boreal forest ecosystems of North America. *Annual Review of Entolmology*, **43**, 107–127.

McInnes, P.F., Naiman, R.J., Pastor, J. & Cohen, Y. (1992). Effects of moose browsing on vegetation and litter of the boreal forest, Isle Royale, Michigan, USA. *Ecology*, **73**, 2059–2075.

Meyer, H.A. (1952). Structure, growth, and drain in balanced uneven-aged forests. *Journal of Forestry*, **50**, 85–92.

Meyer, W.H. (1930). *Diameter Distribution Series in Even-aged Forest Stands.* Yale University School of Forestry Bulletin No. 28.

Michigan State University (1981). *Soil Associations of Michigan.* East Lansing, Michigan: Cooperative Extension Service and Agricultural Experiment Station.

Miller, R.K. (1978). The Keetch-Byram drought index and three fires in Upper Michigan. In *5th National Conference on Fire and Forest Meteorology*, pp. 63–67. Boston, Massachusetts, USA: American Meteorological Society.

Minnesota Biological Survey (1995). *Maple-basswood Forest of Hennepin County: A Threatened Habitat.* St. Paul, Minnesota: Minnesota Department of Natural Resources.

Mladenoff, D.J. (1987). Dynamics of nitrogen mineralization and nitrification in hemlock and hardwood treefall gaps. *Ecology*, **68**, 1171–1180.

Morey, H.F. (1936). Age–size relationships of the Heart's Content, a virgin forest in northwestern Pennsylvania. *Ecology*, **17**, 251–257.

Murie, A. (1934). *The Moose of Isle Royale.* University of Michigan Museum of Zoology Miscellaneous Publication 25.

Muller, R. (1966). Snowbelts of the Great Lakes. *Weatherwise*, **19**, 250–251.

Nakashizuka, T. & Iida, I. (1995). Composition, dynamics and disturbance regime of temperate deciduous forests of Monsoon Asia. *Vegetatio*, **121**, 23–30.

National Oceanic and Atmospheric Administration (1980). *Climatological Data National Summary 1980.* Asheville, North Carolina, USA: Environmental Data Service, National Climatic Center.

Noble, M.G., DeBoer, L.K., Johnson, K.L., Coffin, B.A., Fellows, L.G. & Christensen, N.A. (1977). Quantitative relationships among some *Pinus banksiana – Picea mariana* forests subjected to wildfire and postlogging treatments. *Canadian Journal of Forest Research*, **7**, 368–377.

Nowacki, G.J. & Abrams, M.D. (1997). Radial-growth averaging criteria for reconstructing disturbance histories from presettlement-origin oaks. *Ecological Monographs*, **67**, 225–249.

Noy-Meir, I. (1975). Stability of grazing systems: an application of predator graphs. *Journal of Ecology*, **63**, 459–481.

Ohkubo, T., Kaji, M. & Hamaya, T. (1988). Structure of primary Japanese beech (*Fagus japonica* Maxim.) Forests in the Chichibu Mountains, central Japan, with species reference to regeneration processes. *Ecological Research*, **3**, 101–116.

Ohmann, L.F. & Grigal, D.F. (1981). Contrasting vegetation responses following two forest fires in northeastern Minnesota. *American Midland Naturalist*, **106**, 54–64.

Ohmann, L.F. & Ream, R.R. (1971). *Wilderness Ecology: Virgin Plant Communities of the Boundary Waters Canoe Area*. St. Paul, Minnesota: U.S. Department of Agriculture, Forest Service, North Central Forest Experiment Station.

Oliver, C.D. (1978). *The Development of Northern Red Oak in Mixed Stands in Central New England*. Yale University School of Forestry and Environmental Studies, Bulletin No. 91.

Oliver, C.D. (1981). Forest development in North America following major disturbances. *Forest Ecology and Management*, **3**, 153–168.

Oliver, C.D. & Larson, B.C. (1996). *Forest Stand Dynamics*. Update edition. New York: John Wiley and Sons.

Oliver, C.D. & Stephens, E.P. (1977). Reconstruction of a mixed-species forest in central New England. *Ecology*, **58**, 562–572.

Ostfeld, R.S., Manson, R.H. & Canham, C.D. (1997). Effects of rodents on survival of tree seeds and seedlings invading old fields. *Ecology*, **78**, 1531–1542.

Pacala, S.W., Canham, C.D., Saponara, J., Silander, J.A. Jr., Kobe, R.K. & Ribbens, E. (1996). Forest models defined by field measurements: estimation, error analysis and dynamics. *Ecological Monographs*, **66**, 1–43.

Pastor, J., Aber, J.D., McClaugherty, C.A. & Mellilo, J.M. (1984). Aboveground production and N and P cycling along a nitrogen mineralization gradient on Blackhawk Island, Wisconsin. *Ecology*, **65**, 256–268.

Pastor, J., Gardner, R.H., Dale, V.H. & Post, W.M. (1987). Successional changes in nitrogen availability as a potential factor contributing to spruce declines in boreal North America. *Canadian Journal of Forest Research*, **17**, 1394–1400.

Payandeh, B. & Ek, A.R. (1971). Observations on spatial distribution and the relative precision of systematic sampling. *Canadian Journal of Forest Research*, **1**, 216–222.

Payette, S., Morneau, C., Sirois, L. & Desponts, M. (1989). Recent fire history of the northern Quebec biomes. *Ecology*, **70**, 656–673.

Peet, R.K. (1992). Community structure and ecosystem function. In *Plant Succession: Theory and Prediction,* ed. D.C. Glenn-Lewin, R.K. Peet, & T.T. Veblen, pp. 103–151. London: Chapman and Hall.

Peterson, C.J. & Carson, W.P. (1996). Generalizing forest regeneration models: the dependence of propagule availability on disturbance history and stand size. *Canadian Journal of Forest Research*, **26**, 45–52.

Peterson, C.J. & Pickett, S.T.A. (1995). Forest reorganization: a case study in an old-growth forest catastrophic blowdown. *Ecology*, **76**, 763–774.

Petty, J.A. & Worrell, R. (1981). Stability of coniferous tree stems in relation to damage by snow. *Forestry*, **54**, 115–128.

Phillips, D.W. & McCulloch, J.A.W. (1972). *The Climate of the Great Lakes Basin*.

Climatological Studies No. 20. Toronto: Environment Canada, Atmospheric Environment Service.

Platt, W.J., Evans, G.W. & Rathbun, S.L. (1988). The population dynamics of a long-lived conifer (*Pinus palustris*). *The American Naturalist*, **131**, 491–525.

Ponge, J-F., André, J, Zackrisson, O., Bernier, N., Nilsson, M-C. & Gallet, C. (1998). The forest regeneration puzzle, biological mechanisms in humus layer and forest vegetation dynamics. *BioScience*, **48**, 523–530.

Poston, T. & Stewart, I. (1978). *Catastrophe Theory and its Applications*. London: Pitman.

Poulson, T. & Platt, W.J. (1996). Replacement patterns of beech and sugar maple in Warren Woods, Michigan. *Ecology*, **77**, 1234–1253.

Raup, H.M. (1957). Vegetational adjustment to the instability of the site. In *Proceedings and Papers of the 6th Technical Meeting of the International Union for the Protection of Nature*, pp. 36–48. Edinburgh, Scotland.

Raup, H.M. (1981). Physical disturbance in the life of plants. In *Biotic Crises in Ecological and Evolutionary Time*, pp. 39–52. New York: Academic Press.

Rebertus, A.J., Kitzberger, T., Veblen, T.T. & Roovers, L.M. (1997). Blowdown history and landscape patterns in the Andes of Tierra del Fuego, Argentina. *Ecology*, **78**, 678–692.

Romme, W.H. (1982). Fire and landscape diversity in subalpine forests of Yellowstone National Park. *Ecological Monographs*, **52**, 199–221.

Romme, W.H. & Knight, D.H. (1981). Fire frequency and subalpine forest succession along a topographic gradient in Wyoming. *Ecology*, **62**, 319–326.

Romme, W.H., Everham, E.H., Frelich, L.E., Moritz, M.A. & Sparks, R.E. (1998). Are large, infrequent disturbances qualitatively different from small, frequent disturbances? *Ecosystems*, **1**, 524–534.

Roth, F. (1898). *On the Forestry Conditions of Northern Wisconsin*. Madison, Wisconsin: Wisconsin Geological and Natural History Survey Bulletin, 1, Economics Series, 1.

Runkle, J.R. (1981). Gap regeneration in some old-growth forests of the eastern United States. *Ecology*, **62**, 1041–1051.

Runkle, J.R. (1982). Patterns of disturbance in some old-growth mesic forests of eastern North America. *Ecology*, **63**, 1533–1546.

Runkle, J.R. (1998). Changes in southern Appalachian canopy tree gaps sampled thrice. *Ecology*, **79**, 1768–1780.

Schmitz, O.J. & Sinclair, A.R.E. (1997). Rethinking the role of deer in forest ecosystem dynamics. In *The Science of Overabundance: Deer Ecology and Population Management*, ed. W.J. McShea, H.B. Underwood, & J.H. Rappole, pp. 201–223. Washington, D.C.: Smithsonian Institution Press.

Schroeder, M.J. & Buck, C.C. (1970). *Fire Weather*. U.S. Department of Agriculture, Forest Service, Agriculture Handbook 360.

Scott, M.L. & Murphy, P.G. (1987). Regeneration patterns of northern white cedar, an old-growth forest dominant. *American Midland Naturalist*, **117**, 10–16.

Shugart, H.H. (1984). *A Theory of Forest Dynamics*. New York: Springer-Verlag.

Simard, M.J., Bergeron, Y. & Sirois, L. (1998). Conifer seedling recruitment in a southern Canadian boreal forest: the importance of substrate. *Journal of Vegetation Science*, **9**, 575–582.

Singer, M.T. & Lorimer, C.G. (1997). Crown release as a potential old-growth restoration approach in northern hardwoods. *Canadian Journal of Forest Research*, **27**, 1222–1232.

Smith, T. & Huston, M. (1989). A theory of the spatial and temporal dynamics of plant communities. *Vegetatio*, **83**, 49–69.

Snedecor, G.W. & Cochran, W.G. (1980). *Statistical Methods*, 7th edition. Ames, Iowa: University of Iowa Press.

Sokal, R.R. & Oden, N.L. (1978). Spatial autocorrelation in biology. I. Methodology. *Biological Journal of the Linnean Society*, **10**, 199–228.

Spurr, S.H. (1954). The forests of Itasca in the nineteenth century as related to fire. *Ecology*, **35**, 21–25.

Stark, J.M. & Hart, S.C. (1997). High rates of nitrification and nitrate turnover in undisturbed coniferous forests. *Nature*, **385**, 61–64.

Stearns, F.W. (1949). Ninety years of change in a northern hardwood forest in Wisconsin. *Ecology*, **30**, 350–358.

Stearns, F.W. & Guntenspergen, G.R. (1987a). *Major forest types of the Lake States.* Map, prepared for *The Conservation Foundation Lake States Governor's Conferences on Forestry*, Milwaukee, Wisconsin: University of Wisconsin Cartographic Laboratory.

Stearns, F.W. & Guntenspergen, G.R. (1987b). *Presettlement forests of the Lake States.* Map, prepared for *The Conservation Foundation Lake States Governor's Conferences on Forestry*, Milwaukee, Wisconsin, USA: University of Wisconsin Cartographic Laboratory.

Stein, S.J. (1988). Explanations of the imbalanced age structure and scattered distribution of ponderosa pine within a high-elevation mixed coniferous forest. *Forest Ecology and Management*, **25**, 139–153.

Stephens, E.P. (1956). The uprooting of trees: a forest process. *Soil Science Society of America Proceedings*, **20**, 113–116.

Stokes, M.A. (1981). The dendrochronology of fire history. In *Proceedings of the Fire History Workshop*, U.S. Department of Agriculture, Forest Service, General Technical Report RM-81.

Stoeckeler, J.H. & Arbogast, G. (1955). *Forest Management Lessons from a 1949 Windstorm in Northern Wisconsin and Upper Michigan.* U.S. Department of Agriculture, Forest Service, Lake States Forest Experiment Station Paper No. 34.

Stone, D.M. (1975). *Fertilizing and Thinning Northern Hardwoods in the Lake States.* U.S. Department of Agriculture, Forest Service, North Central Forest Experiment Station. Station Paper NC-141.

Stroempl, G. (1983). *Growth Response of Basswood and Sugar Maple to an Intermediate Cutting.* Ontario Ministry of Natural Resources, Forest Research Paper No. 107.

Swain, A.M. (1973). A history of fire and vegetation in northeastern Minnesota as recorded in lake sediments. *Quaternary Research*, **3**, 383–396.

Swain, A.M. (1978). Environmental changes during the last 2000 years in north-central Wisconsin: analysis of pollen, charcoal, and seeds from varved lake sediments. *Quaternary Research*, **10**, 55–68.

Tester, J., Starfield, A. & Frelich, L.E. (1997). Modeling for ecosystem management in Minnesota pine forests. *Biological Conservation*, **80**, 313–324.

Thom, C.H.S. (1963). Tornado probabilities. *Monthly Weather Review*, **91**, 730–736.

Thom, R. (1975). *Structural Stability and Morphogenesis*. Reading, Massachusetts: Benjamin.

Trewartha, G.T. (1961). *The Earth's Problem Climates*. Madison, Wisconsin: University of Wisconsin Press.

Tubbs, C.H. (1977). Age structure of a northern hardwood selection forest. *Journal of Forestry*, **75**, 22–24.

Turner, M.G., Romme, W.H., Gardner, R.H. & Hargrove, W.W. (1997). Effects of patch size and fire pattern on early post-fire succession on the Yellowstone Plateau. *Ecological Monographs*, **67**, 411–433.

United States Department of Agriculture (1964). *Winds over Wildlands – a Guide for Forest Management*. Agriculture Handbook No. 272. Washington D.C.: U.S. Government Printing Office.

Vanclay, J.K. (1992). Modelling regeneration and recruitment in a tropical rain forest. *Canadian Journal of Forest Research*, **22**, 1235–1248.

Van Wagner, C.E. (1977). Conditions for the start and spread of crown fire. *Canadian Journal of Forest Research*, **7**, 23–34.

Van Wagner, C.E. (1978). Age-class distribution and the forest fire cycle. *Canadian Journal of Forest Research*, **8**, 220–227.

Veblen, T.T. (1986). Treefalls and the coexistence of conifers in subalpine forests of the central Rockies. *Ecology*, **67**, 644–649.

Veblen, T.T. & Stewart, G.H. (1982). On the conifer regeneration gap in New Zealand: the dynamics of *Libocedrus bidwillii* stands on South Island. *Journal of Ecology*, **70**, 413–436.

Visher, S.S. (1954). *Climatic Atlas of the United States*. Cambridge, Massachusetts: Harvard University Press.

Watson, A. (1983). Eighteenth century deer numbers and pine regeneration near Braemar, Scotland. *Biological Conservation*, **25**, 289–305.

Watt, A.S. (1947). Pattern and process in the plant community. *Journal of Ecology*, **35**, 1–22.

Webb, T., III. (1987). The appearance and disappearance of major vegetational assemblages: long-term vegetational dynamics in eastern North America. *Vegetatio*, **69**, 177–187.

Weber, M.G. & Stocks, B.J. (1998). Forest fires and sustainability in the boreal forests of Canada. *Ambio*, **27**, 545–550.

Wells, R.W. (1968). *Fire at Peshtigo*. Englewood Cliffs, New Jersey: Prentice Hall.

West, D.C., Shugart, H.H. & Botkin, D.B. (eds.) (1981). *Forest Succession: Concepts and Applications*. New York: Springer-Verlag.

Whelan, R.J. (1995). *The Ecology of Fire*. Cambridge: Cambridge University Press.

Whitney, G.G. (1986). Relation of Michigan's presettlement pine forests to substrate and disturbance history. *Ecology*, **67**, 1548–1559.

Whitney, G.G. (1987). An ecological history of the Great Lakes forest of Michigan. *Journal of Ecology*, **75**, 667–684.

Whitney, L.F. (1977). Relationship of the subtropical jet stream to severe local storms. *Monthly Weather Review*, **105**, 398–412.

Whittaker, L.M. & Horn, L.H. (1982). *Atlas of Northern Hemisphere Extratropical Cyclone Activity, 1958–1977*. Madison, Wisconsin: University of Wisconsin Department of Meteorology.

Wilson, J.B. & Agnew, A.D.Q. (1992). Positive feedback switches in plant communities. *Advances in Ecological Research*, **23**, 263–336.

Yamamoto, S-I. (1989). Gap dynamics in climax *Fagus crenata* forests. *Botanical Magazine Tokyo*, **102**, 93–114.

Yarie, J. (1981). Forest fire cycles and life tables: a case study from interior Alaska. *Canadian Journal of Forest Research*, **11**, 554–562.

Zackrisson, O., Nilsson, M.C., Dahlberg, A. & Jäderlind, A. (1997). Interference mechanisms in conifer-*Ericaceae*-feathermoss. *Oikos*, **78**, 209–220.

Zeeman, E.C. (1976). Catastrophe theory. *Scientific American*, **234**, 65–83.

Appendix 1

Common tree species of the Great Lakes Region

Scientific name	Common name(s)
Abies balsamea	Balsam fir
Acer rubrum	Red maple
Acer saccharinum	Silver maple
Acer saccharum	Sugar maple
Betula papyrifera	Paper birch
Betula alleghaniensis	Yellow birch
Carya cordiformis	Yellowbud hickory
Carya ovata	Shagbark hickory
Fagus grandifolia	Beech
Fraxinus americana	White ash
Fraxinus nigra	Black ash
Fraxinus pennsylvanica	Green ash
Larix laricina	Tamarack or larch
Ostrya virginiana	Ironwood
Picea glauca	White spruce
Picea mariana	Black spruce
Pinus banksiana	Jack pine
Pinus resinosa	Red pine
Pinus strobus	White pine or eastern white pine
Populus balsamifera	Balsam poplar
Populus deltoides	Cottonwood
Populus grandidentata	Big tooth aspen
Populus tremuloides	Quaking aspen or aspen
Quercus alba	White oak
Quercus bicolor	Swamp white oak
Quercus ellipsoidalis	Northern pin oak or pin oak
Quercus macrocarpa	Bur oak
Quercus rubra	Northern red oak or red oak
Quercus velutina	Black oak
Thuja occidentalis	Northern white cedar or white cedar

Common tree species of the Great Lakes Region (*cont.*)

Scientific name	Common name(s)
Tilia americana	Basswood
Tsuga canadensis	Hemlock (eastern)
Ulmus americana	American elm
Ulmus rubra	Slippery elm

Index

Page numbers in *italics* refer to photographs.